T0365992

Mathematical Modeling and Simulation

Mathematical Modeling and Simulation
Case Studies on Drilling Operations in the Ore Mining Industry

Edited by

Dr. P. N. Belkhode
Assistant Professor
Laxminarayan Institute of Technology, Nagpur

Dr. J. P. Modak
Emeritus Professor, Ph.D (Mechanism), D.Sc (HPFM)
Ex-Professor, Visvesvaraya National Institute of Technology, Nagpur
Advisor Technical
J D College of Engineering and Management, Nagpur

Dr. V. Vidyasagar
Director
National Power Training Institute, Bangalore

Dr. P. B. Maheshwary
Director
J D College of Engineering and Management, Nagpur

CRC Press
Taylor & Francis Group
Boca Raton London New York

CRC Press is an imprint of the
Taylor & Francis Group, an **informa** business

First edition published 2022
by CRC Press
6000 Broken Sound Parkway NW, Suite 300, Boca Raton, FL 33487-2742

and by CRC Press
2 Park Square, Milton Park, Abingdon, Oxon, OX14 4RN

Library of Congress Cataloging-in-Publication Data
Names: Belkhode, Pramod, editor.
Title: Mathematical modelling and simulation: case studies on drilling operations in the ore mining industry / edited by Dr. P.N. Belkhode, Assistant Professor Laxminarayan Institute of Technology, Nagpur, Dr. J. P. Modak, Emeritus Professor, Ph.D (Mechanism), D.Sc (HPFM), Ex-Professor, Visvesvaraya National Institute of Technology, Nagpur, Advisor Technical, J D College of Engineering and Management, Nagpur, Dr. V. Vidyasagar, Director Power Systems Training Institute, Bangalor, Dr. P. B. Maheshwary, Director, J D College of Engineering and Management, Nagpur.
Description: First edition. | Boca Raton, FL: CRC Press, 2021. |
Includes bibliographical references and index.
Identifiers: LCCN 2021007997 (print) | LCCN 2021007998 (ebook) |
ISBN 9780367676353 (hardback) | ISBN 9780367676360 (paperback) |
ISBN 9781003132127 (ebook)
Subjects: LCSH: Boring—Case studies. | Mining engineering—Mathematics. |
Human-machine systems—Mathematical models.
Classification: LCC TN281 .M325 2021 (print) | LCC TN281 (ebook) |
DDC 622/.24015118—dc23
LC record available at https://lccn.loc.gov/2021007997
LC ebook record available at https://lccn.loc.gov/2021007998

ISBN: 978-0-367-67635-3 (hbk)
ISBN: 978-0-367-67636-0 (pbk)
ISBN: 978-1-003-13212-7 (ebk)

Typeset in Times
by codeMantra

Contents

Preface

Manless factories are being developed in countries worldwide. Automation for various manufacturing activities has been developed to such an extent that in such manless factories, all operations related to stores, manufacturing operations such as shape changing of raw material at various stations, inter-station quality assurance in assembly operation, packing operation, raw material, and/or semi-finished product, handling from one station to another, are performed by mechanical hardware electrical cum electronic communication, and synchronization of these systems is being controlled by proper computer programming.

All manufacturing operations or service station activities are not yet completely autonomous. Despite the development of manless factories there are some sectors where human-operated activities are still taking place such as in rural and underdeveloped areas which perform operations using man–machine systems.

In many of such man–machine systems, the aspect of posture adopted by operator, design of tools being handled, atmospheric working conditions, psychological status of operators, condition of tools being used, psychological harmony of teammates do not remain up to the desired performance level. This situation adversely affects (i) productivity, (ii) quality of process or product, and (iii) human energy input of all teammates. There are some manufacturing-related activities such as (i) human operation of physical tools, (ii) manual inter-station material handling, (iii) monitoring of quality of product or process at intermediate manufacturing operations, (iv) assembly operation, and (v) packing still executed through man–machine systems.

Unfortunately, no appropriate quantitative approaches have been developed to decide which aspects of the present methods of performing an activity are weak or strong. In other words, no approaches have been developed for the optimization of man–machine systems because no approaches are established for mathematical modeling of man–machine systems. It is well known that having a mathematical model or simulation of any activity including man–machine system is a fundamental requisite for deciding most optimum input or causes of any activity to advance the desired effects.

The main focus of this book is as follows: how to establish a mathematical model or simulation of man–machine systems so that optimum requirements realize the targeted effects such as (i) maximum production, (ii) best quality of product/process, (iii) minimum human energy inputs realized by adopting the techniques of multivariate constrained optimization of man–machine system.

This has necessitated the evaluation and adoption of the concept of field data-based modeling of man–machine system. This concept was invented by one of this book's authors, professor J. P. Modak, while guiding around 30 doctoral research candidates and establishing mathematical simulation of (i) human assembly operation of plastic luggage manufacturing industry, (ii) small capacity electric motor manufacturing industry, (iii) pedal sewing machine, (iv) traffic noise influence, (v) bicycle peddling and human peddle-driven machine, (vi) maintenance including over handling of narrow gauge train locomotive, (vii) automobile service station, (viii) heavy-duty metal stamping presses, (ix) face drilling operation in underground mines, (x) civil construction activities such as constructing for casting of columns, slabs, (xi) complete functioning of medium duty plastic industry, (xii) manual operation of small capacity flour mills, and so on.

Field data-based modeling approach is applied for simulation of man–machine system—Drilling Operations in the Ore Mining Industries. After getting operational data, the mathematical model is formed followed by optimization, which results in the suggestion in improvement of concerned man–machine system. This is the whole matter emphasized in this book. The book is fundamentally useful for industry personnel for improving the performance of present man–machine system, thereby earning more profit.

MATLAB® is a registered trademark of The MathWorks, Inc. For product information, please contact:

The MathWorks, Inc.
3 Apple Hill Drive
Natick, MA 01760-2098 USA
Tel: 508-647-7000
Fax: 508-647-7001
E-mail: info@mathworks.com
Web: www.mathworks.com

Editors

Dr. P. N. Belkhode did his doctoral research degree in Mechanical Engineering from Rashtrasant Tukadoji Maharaj Nagpur University, Nagpur. He has been an Assistant Professor at Laxminarayan Institute of Technology since 2009. He has published more than 30 research papers in international journals and delivered more than 20 guest lectures on various topics. Four students have been awarded doctoral degrees under his supervision.

The highly esteemed and illustrious **Dr. J. P. Modak** has been serving as the Professor of Mechanical Engineering at J. D. College of Engineering, Nagpur. He served as the Professor in Mechanical Engineering Department, Visvesvaraya National Institute of Technology (VNIT, Nagpur) formerly VRCE till 2002. He worked as Dean of Research and Development at LTJSS, Nagpur from 2002 to 2018. Dr. Modak holds B.E. in Electrical and Mechanical Engineering, ME by research in the area of Stress Analysis, Ph.D in Mechanism and D.Sc in the concept of the Human Powered Flywheel Motor from RTM Nagpur University. With a keen interest in the areas of research of automation, the theory of experimentation, ergonomics, man–machine system, applied robotics, rotor dynamics and kinematics and dynamics of mechanism, Dr. Modak has contributed more than 800 research paper publications, including more than 100 journals of international repute. He has won the national award for the "Best Research Paper" thrice during his eminent career. His genius has been instrumental in developing various products and applications in the field of Mechanical Engineering. He has turned out around 90 Ph.D students and 12 Post graduate degrees by research candidate.

Dr. V. Vidyasagar presently is Director at National Power Training Institute. He earned his Doctorate in Mechanical Engineering from Rashtrasant Tukadoji Maharaj Nagpur University, Nagpur (MS), Master in Thermal Engineering from Indian Institute of Technology, New Delhi and Graduation in Mechanical Engineering from Nagarjuna University A.P. (India). Dr. Vidyasagar served in Central Electricity Authority (GOI) in Investigation and Appraisal of Thermal Power Plants from techno-economic perspective for about a decade. He also conducted training in operations, efficiency and performance of thermal power plant for about 25 years in the National Power Training Institute.

Dr. P. B. Maheshwary did his double doctoral research degree in two different specializations (Machine Design and Thermal Engineering) of Mechanical Engineering. An accomplished teaching professional with more than 3 decades of teaching experience and 10 years of research experience, he has published more than 20 research papers in SCI and Scopus (A+) indexed international journals. With in-depth administrative experience as the Director of an Educational Institute since 2008, Dr. Maheshwary is a self-driven, result-oriented person with flexibility and ability to connect to all levels in an organization. Continuous self-development and continuous learning has added to his knowledge and is an asset to the Institution with which he is associated.

Contributors

Pratibha Agrawal
Laxminarayan Institute of Technology

Vinod Ganvir
Laxminarayan Institute of Technology

Manoj Meshram
Laxminarayan Institute of Technology

Sarika Modak
Priyadarshini College of Engineering

Sagar Shelare
Priyadarshini College of Engineering

Anand Shende
Laxminarayan Institute of Technology

1 Man–Machine System

Pramod Belkhode
Laxminarayan Institute of Technology

J. P. Modak
Visvesvaraya National Institute of Technology and
JD College of Engineering and Management

V. Vidyasagar
Power Systems Training Institute

P. B. Maheshwary
JD College of Engineering and Management

CONTENTS

1.1 INTRODUCTION

Most of the industrial activities are executed manually due to the limitations of mechanization such as technological- and cost-oriented. Industrial activities such as maintenance operations, mining operations, loading and unloading operations on process machines, and similar other operations are manually performed. The management attitude is conservative and traditional with spontaneous judgment-based decision-making predominance. Hence, most of the operations are carried out manually. Operators are working with different types of machine tools and process machines under different environmental conditions. The ergonomic design of the workstation varies suitable for the operators with different constraints. According to academicians, ergonomics deals in an integrated way with the man, his working environment, tools, materials, and process [1]. Work can be completed efficiently with human comfort if designed based on the ergonomic principles. If ergonomic principles are not followed for the design of man–machine systems, then it results in low efficiency, poor health, and rise in the accidents of the man–machine system.

Various operational factors are identified to optimize the productivity and conserving human energy needed for operations. There are many approaches to develop/upgrade industrial activities such as method study (motion study), work measurement

1

(time study), productivity, and so on. Field data-based modeling approach is proposed to study man–machine system.

1.2 CAUSE-AND-EFFECT RELATIONSHIPS

Formulation of logic-based model correlating causes and effects is not possible for these types of complex phenomena. The only approach appropriate for this type of study of phenomenon, that is, man–machine system, is by field data-based modeling. Field data-based model correlates the inputs or causes, in other words, the outputs of such activity by formulating the quantitative mathematical modeling. The indices of the causes of the model, that is, mathematical model, indicate the most influencing inputs. Such correlation indicates the deficiency and the strength of the man–machine system, which helps to improve the performance of the system. Hence, for improving the system/activity performance, it is essential to form such analytical cause–effect relationships conceptualized as field data-based models.

1.3 ERGONOMICS

Ergonomics is the subject dealing with the interaction of the human operator and the physical system of works or system. The performance of man–machine system is optimized by designing the system using ergonomic principles, data, and methods. Ergonomists contribute to the design and evaluation of the system based on the type of job, environment, product, and task to be performed to make them compatible. These principles are used for enhancing safety, reducing fatigue, and increasing comfort with improved job satisfaction while enhancing effectiveness, that is, productivity in carrying out the tasks.

1.4 ANTHROPOMETRY

Anthropometry deals with the measurement of the size and proportions of the human body and parameters such as reach and visual-range capabilities. The application of anthropometry in the design of tools/equipment is to incorporate the relevant human dimensions, aiming to accommodate at least 90% of potential users, considering static factors (height, weight, shoulder breadth, etc.) and dynamic (functional) factors (body movement, distance reach, and movement pattern). Anthropometry is the branch of ergonomics that deals with different human body dimensions of operator accommodated by providing adjustability in the machine used for various operations. Hence, the anthropometric data of operator have been collected.

Postural discomfort is experienced by the operator because of muscular discomfort required to maintain the body posture during work. The interaction of operator with machine process during operation leads to ugly postures; therefore, it is felt necessary to study the specifications of machine from postural comfort's view. In the physiological response study, human energy consumed while performing the task using tools of different designs is recorded. In general, the selection of tool is made on the basis of the constructional features of the machine, tools, and operating and environmental conditions. These form the independent variables of the activities. The physiological cost (dependent/response variable) incurred in the operation is recorded for different conditions of these independent variables. Productivity of operation is considered as other dependent/response variable. These variables are also recorded to study the effects of independent variables on the quality of operation.

Hilbert [2] suggested the experimentation theory to know the output of any activity in terms of various inputs of any phenomenon. It is felt that such an approach is not yet seen toward correctly understanding the operation performed by human being. This approach finally establishes a field data-based model for the phenomenon. Various inputs in the industrial activity are as follows: (i) body specifications of the operator (namely, the anthropometric measurements), (ii) specifications of machine, (iii) specifications of tools, (iv) other process-related parameters, and (v) specifications of environmental factors such as ambient temperature, humidity, air circulation, and so on, at the place of work. The response variables of the phenomenon are as follows: (i) time of operation, (ii) productivity, (iii) human energy consumed during the operation, (iv) quality of operation, and so on. A quantitative relationship is established among the responses and inputs. The inputs and the corresponding responses are measured. Such quantitative relationships are known as mathematical models. Two types of models are established, namely, the model using the concept of least-square multiple regression curve (hereafter referred to as mathematical model) and the other is "Artificial Neural Network"-based model. The interest of the operator lies in arranging inputs to obtain targeted responses. Once the models are formed, they are optimized using the optimization technique [32]. The optimum conditions at which the independent variables should be set for the maximum productivity, minimum human energy expenditure, and accepted quality are deduced. From the analysis of models, the intensity of influence of various independent variables on the dependent variables and the nature of relationship between independent and dependent variables are determined. Finally, some important conclusions are drawn based on the analysis of models.

1.5 APPROACH TO FORMULATE

This book mainly aims to explain the approach to formulate the mathematical model for the man–machine system. To form the mathematical model, the most critical industrial activities - mining industry is identified and studied. Mining industries involve operations such as face drilling, loading operation, unloading operation, and roof bolting. The formulation of a mathematical model for manual work such as face drilling at underground mines is selected as the case study. This work describes the present method of carrying out the face drilling operations. In the present method, the productivity is less, and human energy requirement is substantial. The variables related to the face drilling operation in the mining are identified to enhance the productivity with minimum human energy. In this book, the approximate generalized mathematical models have been established by applying the concepts of theories of experimentation [2] for the face drilling operations in underground mines. The general procedure adopted is as follows.

1. Review the existing literature on industrial operations: A general overview in relation to ergonomic aspects in which the interaction between man–machine system.

2. Study of existing workstation including man–machine system. This study includes maintenance, schedule, problems, functions, purpose, benefits, and so on.
3. Possibility of formulation of model for improving industrial activity: The industrial activity is very difficult to plan and involves high costs. Therefore, there is a need for developing model, which helps to reduce human energy and repair time. Data are collected based on the sequence of industrial activity by direct measurement. From this, data input and output variables are decided, and the model is formed by forming a dimensionless equation using regression analysis.
4. Possibility to validate output and model of the system: the aim is to find out the utility and the effectiveness of model. The effectiveness of model is decided by ANN simulation, sensitivity analysis, and optimization technique.

2 Concept of Field Data Data-Based Modeling

Pramod Belkhode
Laxminarayan Institute of Technology

Sarika Modak
Priyadarshini College of Engineering

V. Vidyasagar
Power Systems Training Institute

P. B. Maheshwary
JD College of Engineering and Management

In our life, we come across many more activities. These activities have some environmental systems in which these activities occur. The environment or system can be specified in terms of its parameters – some of which are always constant in their magnitudes, whereas some are variable. The activities are set in action by some parameters, which are considered as causes. These causes interact with parameters of the system; as a result of this interaction, some effects are produced. The above mentioned matter in a diagrammatic form can be presented, as shown in Figure 2.1.

Figure 2.1 shows a rectangular block where activity along with its nature is written, that is, activity may be physical or it may be a combination of human-directed/-operated physical activity stated as man–machine system or it may be an activity

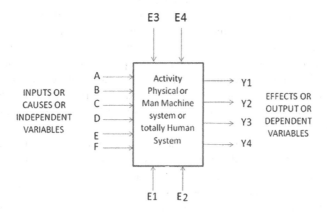

FIGURE 2.1   Block diagrammatic representation of an activity.

mainly dominated by human being stated in the block as "Totally Human System".
The functioning of an activity is influenced by two sets of parameters: the first set
characterizes the features of environment of an activity and the other set characterizes planned parameters or causes, which influence the functioning of the system.

Accordingly, Figure 2.1 shows these parameters. The parameters characterizing
features of an environment are E1, E2, E3, E4, and so on. Some of which are perma-
nently fixed, shown as E1, E2, whereas remaining are time variants, shown as E3, E4
on which one has no control. The planned parameters are known as causes shown
as A, B, C, D, E, and so on, and the effects of an activity are shown as Y1, Y2, Y3,
Y4, and so on.

If one analyzes any activity of society, then one would be able to identify the
causes A to E, and so on, system parameters E1 to E4, and so on, and the effects Y1
to Y4, and so on. This may be treated as qualitative analysis of the societal activity.
This can be demonstrated by one example of everyday life of man–machine system
of our life.

Let us consider a gardener preparing flower beds in a kitchen garden of a conventional house of a slightly upper middle-class family.

Supposing the house owner has to prepare around 6 to 8 flower beds. Each is of
the size 3 m in length, 1 m wide, and 0.5 m in depth. The house owner instructs his
gardener accordingly. Let us say that gardener decides to start the work from specific
day along with his team of 2 to 3 helpers. The tools necessary for this operation are
(i) a Kudali (axe), (ii) a phawada (a spade), and (iii) a soil collector a Ghamela. The
work would take place in a shift of 8 hours, that is, since 8.00 am in the morning till
5.00 pm in the late afternoon with a rest cum lunch break from 12.30 pm to 1.30 pm.
Let us say that it is a team of total three members with one supervisor performing
this task. The first member says that A digs the soil, the second member that B collects the dry soil with spade and puts it in ghamela, and the third member says that C
carries the soil to the place where heap of soil is made.

The planned sequence of working is A dugs for 10 minutes with rest pause for 3
minutes at the end of every 7 minutes. During this rest pause of 3 minutes collects
this duged soil in ghamela and C carries this soil to the heap. Like this, A, B, C workers will work for every 30 minutes in a sequence. During these 3 minutes, the leader
of the team with take down the observations of activity performed in 7 minutes of
digging and 3 minutes of carrying duged soil to heap and measuring the duged quantity in Kgf. The measurements taken are (i) initial pulse rate of A, B, C, (ii) pulse
rate at the end of 7 minutes of digging for A, (iii) pulse rate of B and C at the end of
10th minute, (iv) the measurement of rise in body temperature would also be noted,
(v) the measurement of soil dug in kgf at the end of every 10 minutes will also be
recorded, and (vi) the surface finish of sides of rectangular space created by A will
also be noted.

The complete set of observations of above listed parameters at the end of every
30 minutes would be recorded. Along with this record would also be maintained during the beginning and end of 30 minutes with regard to attitude and enthusiasm of
workers. The specifications of the tools in terms of their geometry, weight, sharpening of digging point, and edge of the spade. At the end of the shift on every day, the
total of all causes and effects will be made. Referring to Figure 2.1 for this operation

cause A may be anthropometric dimensions of operators including their number, B represents experience and qualifications of operators, E represents enthusiasm and attitude of operators may be geometric dimensions of the tools, their weight, and their condition, that is, sharpening of edges and tips represents soil condition at the spot before digging, that is, at the beginning of every 10 minutes represents time of operation, whereas E1 and E2 could be general features of kitchen garden, E3 represents ambient temperature, pressure, and humidity, and E4 represents noise level. E1 and E2 could be constant parameters and E3 and E4 represents extraneous variables of the system, that is, activity under consideration in this case. The responses of the activity could be total soil dug out in every 10 minutes, for example Y1, human energy input in terms of pulse rate or blood pressure rise Y2, and the quality of operation performed Y3.

Thus, in this case, one can say that E1 and E2 are constant parameters of system, whereas E3 and E4 are system extraneous variables. A, B, C, D, E, and F are planned and/or actual unplanned but measured causes or inputs and Y1, Y2, and Y3 could be responses/outputs of the activity.

In the list of abovementioned variables, there are some variables that are very difficult to measure, for example, enthusiasm and attitude of the operator. These are categorized as abstract inputs or causes and sometimes effects, that is, how the worker feels psychologically at the end of every 10 minutes. However, these abstract quantities can be measured using the concept of weight ages. In brief, all causes, constant parameters of the system, some of extraneous variables, and effects can be quantified. In that case, one can arrive at the complete observation table of this activity. In the context of completely physical system, it could have been specified as phenomenal or experimental observations.

Once these observations are ready, one can form the mathematical co-relationships among (i) causes, (ii) effects, and, to some extent, (iii) extraneous variables of the form as far as this activity is concerned as

$$Y1 = K1(A)^{a1}(B)^{b1}(C)^{c1}(D)^{d1}(E1)^{e11}(E2)^{e21} \tag{2.1}$$

$$Y2 = K2(A)^{a2}(B)^{b2}(C)^{c2}(D)^{d2}(E1)^{e12}(E2)^{e22} \tag{2.2}$$

$$Y3 = K3(A)^{a3}(B)^{b3}(C)^{c3}(D)^{d3}(E1)^{e13}(E2)^{e23} \tag{2.3}$$

Equations 2.1 to 2.3 can be formed based on information presented in observation and using the already available mathematical treatments especially matrix algebra. All exponents of Equations 2.1 to 2.3 can be obtained. Quantities K1, K2, and K3 are known as curve-fitting constants. They represent collectively what is not identified as causes logically and/or as a result of action of extraneous variables quantitatively but collectively.

Once these Equations 2.1 to 2.3 are formed, they can be then looked upon as a design tool for planning similar activity in future and not necessarily with the same conditions but could be applicable for variation of all causes and constant parameters of the system within say 80% to 120% range of their variation. Equations 2.1 to 2.3

can be considered as field data-based models the concept first launched by Dr. J. P. Modak, one of the authors of this book Dr. J. P. Modak in the body of knowledge. It is a field data-based model because it is formed based on actual field studies. The approach is strongly advocated for any man–machine system of our life. Accordingly, Dr. Modak has guided around 21 candidates for their Doctoral Research work based on this concept of field data-based modeling of several man–machine systems such as (i) human assembly operation, (ii) sewing machine operations, (iii) agricultural veeding operation, street noise, face drilling in coal mines, pharmaceutical industry, heavy-duty press operation, cotton ginning operation, civil construction activities, factory maintenance jobs, floor mill operation, and plastic industry operation.

Accuracy of models is dependent on magnitude of curve-fitting constants such as K1, K2, and K3. Ideally, if these are numerically 1, then the model rightly simulates the man–machine system. If it is too low, the causes are overestimated, and if it is too high, they the causes are under estimated. This would be decided when to repeat the investigation or to refine the approach in subsequent attempts. The magnitude of exponents of the causes, that is, of quantities on right-hand side of Equations 2.1–2.3 indicates the degree of influences of these causes on the specific response. The ratio of indices indicates the relative influences of two causes, that is, if say exponent of B is higher than that of C, then the interpretation would be the qualification, experience of operator is more influential than that of quality of tools used.

The approach is extremely important for technologists working in industries. Because presently for some industry while designing a new plant for the product similar to the one being manufactured presently, the design approach is normally based on the opinion of individual group member. It goes on like this: at first, meeting of seniors some member may opine saying, "I think it is sophisticated equipment is more influential then the qualities of human resource". The formulated model for functioning of their present setup if available then that model itself will indicate what is more influential for productivity, product quality, production turn over, or getting low maintenance cost.

This approach suggested by Dr Modak because of his very long experience of applying an approach of "Theories of Engineering Experimentation" by H Schenck Jr. for Formulation of Experimental Data-Based Models for many manufacturing process machines energized by human powered flywheel motor (HPFM). [D.Sc Treatise of Dr. J. P. Modak]. These process machines are (i) rectangular-sectioned brick extruders, (ii) keyed sectioned brick extruders, (iii) wood turning, (iv) low head water lifting, (v) Alge formation machine, (vi) chaff cutting, (vii) Fertilizer ingredients mixing, (viii) ballets making, (ix) food grain crushing, (x) concrete ingredients mixing, (xi) electricity generation, and (xii) groundnuts crushing for manufacturing groundnut oil. Dr. Modak and his research scholars applied this approach of experimental data-based modeling to these process machines energized by HPFM and also two subsystems, that is, energy source HPFM itself and its mechanical power transmission.

HPFM stands for Human Powered Flywheel Motor. This comprises (i) a bicycle like peddling system (ii) speed-increasing gear pair with speed rise ratio G = 2.5 to 30 (iii) a fairly big size flywheel about 1 m rim diameter 10.0 cm rim width 2.0 cm rim thickness.

In less than a minute, a young lad between the ages of 20–25 years, with a slim stature and a height of 160–165 cm spines the flywheel to a speed of 800–1000 RPM. This way around 40,000 N-m of energy is stored in the flywheel. Then, a specially designed torsionally flexible clutch (TFC) is engaged, and this flywheel is connected to process units mentioned earlier.

Upon engaging the clutch instantaneous energy and momentum transfer takes place from high-speed spinning flywheel to input shaft of the processor, which results in manufacturing of product meant by that processor, that is, bricks in the case of clay extruder and so forth.

The energy utilization time is in the range of 5 to 15 seconds depending on average process resistance of the process units. This means that the process units requiring about 11.0 hp to 4.0 hp and where the product quality is not getting affected by constantly changing rotating speed of input shaft of the processor can be energized by such a nonconventional energy source, that is, HPFM.

The functional feasibility and economic viability of this novel machine system are established. All the three sub-systems of this machine system, that is, HPFM the energy source, (ii) process units, and (iii) hence the intermediate system, that is, Torsionally Flexible Clutch (TFC) and torque amplification gear pair G operate all the while in a highly transient state.

It became very difficult to evolve logic based design data for such a man–machine system. Hence the only approach left over was to generate experimental data-based design data. In this approach, all the independent quantities should be experimentally varied to collect the response data and based on this collected data formulate the "Experimental data based Mathematical Model" for all such system.

The approach of Experimental Data-Based Modeling is deduced based on application of approach proposed in the book "Theories of Engineering Experimentation" [2].

This approach step wise is as follows:

i. Identify all causes, responses, and extraneous variables of the phenomena. This can be performed by qualitative analysis of phenomena.
ii. Treat causes as inputs/independent variables
iii. Treat responses as outputs/dependent variables
iv. Combine independent variables by dimensional analysis to reduce them in size. This gives independent Π terms
v. Decide test envelops, test points, and test sequence for all independent Π terms
vi. Design and fabricate the experimental setup
vii. Execute experimentation as per its planning
viii. Collect the experimental data. Upon getting experimental data regarding causes/inputs, independent variables, and similar data for responses, formulate the mathematical models. These models can work as design data.

Well all that has been said above is possible for a totally physically system which is operating in a highly transient state that for such systems Logic-based modeling is highly importable. But what about man–machine systems?

For all man–machine systems, the approach, as mentioned earlier, of Preparing Flower Beds is only applicable. One can say that this approach is similar to experimental data-based modeling approach. Yes this is true but the Design of Experimentation Phase of this approach is not possible for man–machine systems. One has to allow the field activities to take place the way it is planned by their administrates.

The only option is carefully select appropriate sites where activities take place for a specific objective but with variation of all inputs/causes/independents variables and hence corresponding variation of effects/outputs/dependent variables.

Upon getting this information form mathematical models. This is conceptualized as field data-based modeling which is meant for all man–machine systems. When it comes to the formulation of the model, there is no other substitute, but to adopt the methodology of experimentation, more suitably the one suggested by Hilbert [2]. This method is applied to a complex totally physical phenomenon for which logic-based modeling is highly improbable. The inputs to the phenomenon, the outputs and the extraneous variables are identified. The inputs are experimentally varied over a broad yet practically possible range and response data is collected. The curve-fitting constant of the model represents collectively extraneous variables, which do affect the phenomena but cannot be measured. Based on the gathered data, the models are formed.

3 Scope of Book

Pramod Belkhode
Laxminarayan Institute of Technology

J. P. Modak
Visvesvaraya National Institute of Technology and
JD College of Engineering and Management

V. Vidyasagar
Power Systems Training Institute

P. B. Maheshwary
JD College of Engineering and Management

CONTENTS

3.1 INTRODUCTION

An exhaustive literature review has been carried out, which has revealed the categorized information on various ergonomic aspects of face drilling operation in the underground mines. Salient findings of various researchers have been compiled. It is now time to critically examine these observations in view of their implementation in practice.

Many researchers have recommended for ergonomic design and mechanization of mining operations. Hence, it is evident that mechanization of mining operation has to be task-oriented.

3.2 PROBLEMS ASSOCIATED WITH FACE DRILLING ACTIVITIES IN UNDERGROUND MINES

Strength characteristics of the underground miners including back, shoulder, arm, sitting leg strength, and standing leg strength are poor when compared with other industrial workers [3–5]. Most studies agree that underground miners are bound to have lower than average aerobic capacity compared with the population norms and with the comparison groups [6–8]. Occasionally, miners perform physical work in vertical space restrictions such that even crawling is not possible [9]. While this is an extreme case, it is not at all uncommon in the mine for the ceiling to be not higher than 1.5 meters. The physiological and biomechanical demands of doing manual work in such an environment are much greater, with the above mentioned constraints [10–15]. Further, they have to work in humid, less airy, poorly illuminated and noisy environment along with whole-body vibrations. Hence, due to the present face drilling method, the productivity is less, and requirement of human energy and time is substantial. The generalized mathematical model has been formulated using the concept of field data-based modeling for the face drilling activities in underground mines. Therefore, the present approach could be replaced with optimized techniques based on the field data-based modeling in which dependent and independent variables of an activity can be compared, and the most effective method for improving the present method of performing operation can be evolved. Hence, it is required to identify the factors influencing the face drilling necessitate formulating the field data-based models (FDBM) for these activities to increase the productivity besides reducing the time required for performing the operation and conserving human energy.

It can be seen from the literature review and cursory survey that the effectiveness of face drilling not only depends on the specifications of drilling machine/process parameters but also on the anthropometric dimensions of the miner including his attitude and aptitude to do work [16–19]. It also depends on specifications of drill rod, speed of drill and penetration rate of drilling machine, ambient temperature, relative humidity, and illumination at the workstation [20]. Therefore, it is decided to study the effects of these inputs (independent variables or causes) on the effects or outputs of dependent variables, that is, time of face drilling, productivity, and human energy consumed in this activity. In various studies carried out in the assessment of mining operations, no researcher has applied the concept of field data-based modeling (Chapter 2). No research has been carried out so far, which quantitatively specifies the influence of an individual independent variable on response variable. This is only possible by establishing a mathematical model. Hence, it has been decided to demonstrate this in this book. By applying this approach systematically, one would get a new insight into optimization of input parameters for face drilling to achieve maximum efficiency or productivity in these operations. Once the model is formulated for any phenomenon, one gets a clear idea about the variation of dependent variables in terms of interaction of various independent variables. Then applying the optimization techniques [32], determination of the optimum conditions for the execution of the activity becomes possible. Hence, it is always felt that any field data research of any man–machine system should ultimately get precipitated to formulation of a model. Hence, in this research, it has been decided to adopt this approach to obtain the optimization condition, which will enhance the productivity of the operation.

A theoretical approach can be adopted if a known logic can be applied correlating various independent and dependent parameters of the system. Though qualitatively, the relationships between the independent and dependent parameters are known based on the available literature, the generalized quantitative relationships are not known many a times. Hence, formulation of a logic-based model, the only option is to formulate the field-data based model.

Field study-based observations with regard to the response of the phenomenon have shown as to how it varies as the inputs to the phenomenon or independent variables are varied. At times, quantitative relationships correlating a response variable

TABLE 3.1
Interaction of the Inputs on the Response Variables

Types of Variables	Inputs	Description
Causes/Inputs/Independent	a	Anthropometric dimensions of the miner
Variables	b	Specifications of drill rod
	c	Specifications of drilling machine/Process parameters
	d	Speed & penetration rate of drilling machine
	e	Ambient temperature and relative humidity
	f	Illumination at the workstation (environment)
Effects/Outputs/Dependent	A	Time of face drilling
Variables	B	Productivity of face drilling
	C	Human energy consumed in face drilling

with a single input is established. In very few investigations, the influence of simultaneous actions of various inputs in order to create the response is the nature of phenomenon [21–22]. However, a quantitative representation of the interaction between inputs and response is seldom seen.

The present work mainly aims to establish the quantitative relationship of the interaction of the inputs on the response variables as shown in Table 3.1 and to optimize them in face drilling operation in underground mines.

Quantitative relationship is established for the interaction of the inputs (viz. a, b, c, d, e, and f) to precipitate the time of face drilling operation, productivity of face drilling, and human energy consumption during the face drilling operation.

The advantage of such quantitative relationships is to ascertain the relative influence of inputs on the responses. This is possible if such a quantitative relationship, that is a mathematical model, is established. Once the model is developed, the exercise of optimization could be taken up subsequently. The interest of miners lies in arranging optimized inputs to get targeted responses.

4 Design of Field Study

Pramod Belkhode
Laxminarayan Institute of Technology

J. P. Modak
Visvesvaraya National Institute of Technology and
JD College of Engineering and Management

V. Vidyasagar
Power Systems Training Institute

Manoj Meshram
Laxminarayan Institute of Technology

CONTENTS

4.1 INTRODUCTION

Evaluation of face drilling activity is a complex phenomenon. There are many factors (such as anthropometric parameters of operator, his quantifications, experience, attitude, enthusiasm on a specific shift of work, specifications of drilling machine, specifications of drill rod, and environmental variables) affecting the performance of face drilling operation [19]. To study man–machine interaction and human fatigue in various mining operations field data-based modeling approach is proposed.

The field studies to be carried for such investigations need proper planning. In general, a large number of variables are involved in such field studies. It is expected that the influence of all the variables and parameters be studied economically without sacrificing the accuracy by reducing the number of variables to few dimensionless terms through the technique of dimensional analysis. The theory of engineering experimentation is a study of scientific phenomenon, which includes the analysis and synthesis of the scientific phenomenon [2]. The study may include theoretical and experimental approaches.

In the theoretical approach, the laws of mechanics and physics apply, which include (i) force balance, (ii) momentum balance, (iii) energy balance, and (iv) quantity balance. Effects of several independent variables on complex process are studied to formulate the process. In the experimental approach, various steps involved in formulating the model for such a complex phenomenon is as below:

a. **Identify the Causes and Effects**: Performing qualitative analysis of process to identify various physical quantities. These are the Causes (Inputs). Experiment is time-consuming and becomes complex if the large number of independent variables is involved. By deducing the dimensional equation for the phenomenon, the number of variables is reduced.

b. **Perform Test Planning**: This involves deciding test envelope, test points, test sequence, and plan of experimentation.

    i.   **Test Envelope**: To decide the range of variation of individual independent π terms.

    ii.   **Test Points**: To decide and specify values of independent π terms at which experimental setup be set during experimentation.

    iii.   **Test Sequence**: To decide the sequence in which the test points be set during experimentation. Sequence may be of ascending order, descending order, or random order. Usually, ascending or descending order depending on the nature of the phenomenon adopted for irreversible experiment, whereas random order is adopted for reversible experiments.

    iv.   **Plan of Experimentation**: In the planning, the decision is reached regarding how to vary the independent variable. Planning may be the classical plan or factorial plan. In a classical plan, only one variable is varied at a time maintaining all other variables at constant level. In a factorial plan, more than one or all independent variables are varied at a time.

c. **Physical Design of an Experimental setup**: Here, it is necessary to work out the physical design of an experimental setup including deciding specifications and procurement of Instrumentation and Experimentation. The next step would be to execute experimentation as per test planning. This will generate experimental data regarding causes (inputs) and effects (responses).

d. **Checking and Rejection of Test Data**: Based on the experimental results, it is necessary to check it for its reliability. The erroneous data are identified and removed from the gathered data for this purpose, and some statistics-based rules are to be adopted.

    e. **Formulation of the Model**: Quantitative relationship is formulated between the dependent and independent π terms of the dimensional equation. This establishes the relationship between outputs (effects) and inputs (causes).

4.2 LIMITATIONS OF ADOPTING EXPERIMENTAL DATA-BASED MODEL FOR FORMULATING A MODEL FOR MAN–MACHINE SYSTEMS

Man–machine system is complex industrial activities. However, in many such systems, test planning part of experimentation approach is not feasible to be adopted. One has to allow the activity to occur as planned for many other considerations. This occurs when one wishes to formulate a model for any activity in underground mining such as face drilling, civil construction activities, human assembly operations, industry manufacturing activities, and so on.

Though qualitatively, the relationships between the dependent and independent parameters are known based on the available literature references, the generalized quantitative relationships are not known sometimes. Whatever quantitative relationships are known, they are about specific anthropometric data and a specific task. The data for mine workers, engaged in mining task is not available. Hence, formulating the quantitative relationship is not possible. Because there is no possibility of formulation of theoretical model (logic-based), one is left with the only alternative of formulating facts based or to be more specific, in this case, field data-based model. Hence, it is proposed to formulate such a model in the present investigation.

The approach adopted for formulating generalized experimental data-based model suggested by Schenck [2], to be more specific field data-based model suggested by Modak [22] has been proposed in the present investigation which involves the following steps:

- Considering any man–machine system as a phenomenon
- Identification of variables or parameters affecting the phenomenon
- Reduction of variables through dimensional analysis
- Selecting sufficient number of cities with variation of causes and extraneous variables
- Executing fieldwork for data collection
- Rejection of absurd data
- Formulation of the model

Based on this purified data, one has to formulate the quantitative relationship between the dependent and independent π terms of the dimensional equation.

4.3 IDENTIFICATION OF CAUSES AND EFFECTS OF AN ACTIVITY

Identification of causes and effects is the first step. These causes and effects vary as time elapses. The identification of dependent and independent variables, as the phenomena occur, is to be performed based on observed qualitative analysis of the phenomenon. These variables are of four types:

1. Independent variables

   Independent variables can be defined as a cause of the activity that influences the activity. This can be changed independently of the other variables of the activity.

2. Dependent variables

   The phenomenal quantity or parameters which change due to the change in the values of the independent variables are called the response or the dependent variables.

3. Extraneous variables

   Any parameter that influences the process, but its magnitude cannot be changed or altered at our wish such as ambient pressure, humidity, temperature, operator-related parameters such as enthusiasm, attitude, and so on.

4. Controlled variables

   Controlled variables are the phenomenal quantities that remain constant all through the duration of activity which are independent variables, but due to practical reasons, they are not alterable. For example, such as acceleration due to gravity.

4.3.1   DIMENSIONAL ANALYSIS

The dimensional analysis was rightly used primarily as an experimental tool to combine any experimental variable into one (or to reduce the number of experimental variables); this technique was then mainly applied in fluid mechanics and heat transfer for almost all the experiments.

   a. Dimensions/Quantities

      There are two types of quantities or dimensions.

      i.   Fundamental Quantities or Fundamental Units

           Mass (M), Length (L), and Time (T) are three constant dimensions. If heat is involved, then temperature (θ) is also taken as fundamental quantity.

      ii. Derived Dimensions (Secondary Quantities)

           If the physical quantity is designed based on some fundamental, then it is quantities known as derived quantities. The derived quantities classified as shown in Tables 4.1 and 4.2.

TABLE 4.1

Types of the Derived Quantities

S.N.	Variables/Constants	Distinguishing Feature of the Physical Quantity	Example
1	Dimensional variables	Have dimension but no physical value	Force, velocity, power
2	Dimensional variables	Neither have dimension nor fixed value	Specific gravity
3	Dimensional constants	Have fixed dimensions an fixed value	Gravitational constant
4	Dimensional constants	Have no dimension but have fixed values	1, 2, 3

TABLE 4.2
Dimensional Formulae of the Derived Quantities

Sr.N.	Physical Quantity	Relation with Other π Quantities	Dimensional Formula	S.I. Unit
1	Area	Length × breadth	$[L] \times [L] = [L^2]$	m^2
2	Volume	Length × breadth × height	$[L] \times [L] \times [L] = [L^3]$	m^3
3	Density	Mass / volume	$[M]/[L^3] = [M^1 L^{-3} T^0]$	kgm^{-3}
4	Velocity	Distance / time	$[L]/[T] = [M^0 L^1 T^{-1}]$	ms^{-1}
5	Acceleration	Velocity / time	$[M^0 L^1 T^{-1}]/[T] = [M^0 L^1 T^{-2}]$	ms^{-2}
6	Force	Mass × acceleration	$[M] \times [M^1 L^1 T^{-2}] = [M^1 L^1 L^{-2}]$	N
7	Momentum	Mass × velocity	$[M] \times [M^0 L^1 T^{-1}] = [M^1 L^1 T^{-1}]$	$kgms^{-1}$
8	Work	Force × distance	$[MLT^{-2}] \times [L] = [ML^2 T^{-2}]$	Nm
9	Power	Work/time	$[ML^2 T^{-2}]/[T] = [ML^2 T^{-3}]$	W
10	Pressure	Force/area	$[MLT^{-2}]/[L^2] = [ML^{-1} T^{-2}]$	Nm^{-2}
11	Kinetic energy	½ × mass × (velocity)²	$[M] \times [MLT^{-1}]^2 = [ML^2 T^{-2}]$	Nm
12	Potential energy	Mass × g × distance	$[M] \times [M^0 LT^{-2}] \times [L] = [ML^2 T^{-2}]$	Nm
13	Impulse	Force × time	$[MLT^{-2}] \times [T] = [MLT^{-1}]$	Ns
14	Torque	Force × distance	$[MLT^{-2}] \times [L] = [ML^2 T^{-2}]$	Nm
15	Stress	Force / area	$[MLT^{-2}]/[L^2] = [ML^{-1} T^{-2}]$	Nm^{-2}
16	Strain	Extension in length / original length	$[L]/[L] = [M^0 L^0 T^0]$	Number
17	Elasticity	Stress / strain	$[ML^{-1} T^{-2}]/[M^0 L^0 T^0] = [ML^{-1} T^{-2}]$	Nm^{-2}
18	Surface tension	Force/length	$[MLT^{-2}]/[L^1] = [ML^0 T^{-2}]$	Nm^{-1}
19	Force constant of spring	Applied force/extension in length	$[MLT^{-2}]/[L^1] = [ML^0 T^{-2}]$	Nm^{-1}
20	Gravitational constant	Force × (distance)²/ (mass)²	$[MLT^{-2}] \times [L^2]/[M^2] = [M^{-1} L^3 T^{-2}]$	$Nm^2 kg^{-2}$
21	Frequency	1/Time period	$1/[T] = [M^0 L^0 T^{-1}]$	s^{-1}
22	Angle	Arc/radius	$[L]/[L] = [M^0 L^0 T^0]$	Rad
23	Angular velocity	Angle/time	$[M^0 L^0 T^0]/[T] = [M^0 L^0 T^{-1}]$	Rad s^{-1}
24	Angular acceleration	Angular velocity/time	$[M^0 L^0 T^{-1}]/[T] = [M^0 L^0 T^{-2}]$	Rad s^{-2}
25	Moment of inertia	Mass × (distance)²	$[M]/[L^2] = [M^1 L^{-2} T^0]$	kgm^2
26	Angular momentum	Moment of inertia × Angular velocity	$[M^1 L^{-2} T^0] \times [M^0 L^0 T^{-1}] = [ML^2 T^{-1}]$	$kgm^2 s^{-1}$
27	Heat	Energy	$[ML^2 T^{-2}]$	J
28	Planck's constant	Energy/frequency	$[ML^2 T^{-2}]/[M^0 L^0 T^{-1}] = [ML^2 T^{-1}]$	Js
29	Velocity gradient	Change in velocity / distance	$[M^0 LT^{-1}] \times [L] = [M^0 L^0 T^{-1}]$	s^{-1}
30	Radius of gyration	Distance	$[M^0 LT^0]$	m

4.3.2 DIMENSIONAL EQUATION

If in an equation containing physical quantity, each quantity is represented by its dimensional formula, the resulting equation is known as dimensional equation.

$$\text{Kinetic energy} = \tfrac{1}{2}\, mv^2$$

Here, m is the mass of the body, v is velocity. Writing the formula for kinetic energy in the dimensional equation form, we have

$$[M] \times \left[MLT^{-1} \right]^2 = \left[ML^2T^{-2} \right]$$

The abovementioned equation is known as the dimensional equation.

In the dimensional analysis of a physical phenomenon, the relationship between the dependent and independent variables is studied in terms of their basic dimensions to obtain information about the functional relationship between the dimensionless parameters that control the phenomenon. There are several methods of reducing the number of dimensionless parameters. The most commonly used methods are Rayleigh's method and Buckingham Pi theorem method. Sometimes, even by observations and preliminary qualitative analysis, dimensional equation can be formed.

4.3.2.1 Rayleigh's Method

Let Y be an independent variable, which depends on x_1, x_2, x_3, x_4, and so on. According to Rayleigh's method, Y is a function of x_1, x_2, x_3, x_4... etc. and mathematically it can be written as

$$Y = f(x_1, x_2, x_3, x_4)$$

The equation can be written as follows:

$$Y = k\, x_1{}^a, x_2{}^b, x_3{}^c, x_4{}^d$$

where k is constant and a, b, c, and d are arbitrarily indices. The values of a, b, c, and d are obtained by comparing the powers of the fundamental dimension on both sides.

4.3.2.2 Buckingham π Theorem Method

The Buckingham π theorem states that if there are m primary dimensions involved in the variables controlling a physical phenomenon, then the phenomenon can be described by (n − m) independent dimensionless groups. This theorem can be used for reducing the number of variables affecting the process. The theorem states that if any equation is dimensionally homogenous, it can be reduced to a relationship among a complete set of dimensionless product. In this method m, number of repeated variables are selected and dimensionless groups obtained by each one of the remaining variables one at a time. Rayleigh's method is also known as the method of repeating variables. Care is needed in selecting the repeating variables. They must have among themselves all the basic dimensions involved in the problem.

1. The dependent variable must not be chosen as a repeating variable.
2. The repeating variables should be chosen in such a way that one variable contains geometric property, another variable contains flow properly, and the third variable contains fluid property.
3. Usually, a length parameter (D or H); a typical velocity V and the fluid density are convenient sets of repeating variables.

Rayleigh's method of dimensional analysis becomes more laborious if the variables are more than the number of fundamental dimension (M.L.T). This difficulty is overcome by using Buckingham π theorem.

Using this principle, modern experiments can substantially improve their working techniques and made them shorter requiring less time without loss of control [23]. Deducing the dimensional equation for a problem reduces the number of variables in the experiments by applying Buckingham π theorem. If we take the product of the π terms, then it will also be a dimensionless number and hence a π term. This idea is used to achieve further reduction in the number of independent π term variables, which further forms few π terms. Attempt is newly made to apply above discussed matter to a Face drilling operation.

4.3.3 FACE DRILLING OPERATION (INDEPENDENT VARIABLES) PARAMETERS

The variables affecting the effectiveness of the phenomenon under consideration are man, interacting physical system (details of workstation) and environmental (conditions) variables. The dependent or the response variables are as follows:

- Time of face drilling operation (Td)
- Productivity of drilling (Pd) i.e. amount of earthen matter removed
- Consumption of human energy in face drilling operation (He)

For better presentation and manipulation of these independent variables in this study, they have been grouped in the following categories.

- Anthropometric features/variables of operator
- Drill rod parameters
- Drilling machine
- Process parameters
- Environmental variables
- Other variables

Dependent and independent variables for the man–machine system of face drilling operation involved in mining industries are present in Table 4.3.

TABLE 4.3
Independent & Dependent Variables—Face Drilling

Sr. No.	Description	Variables	Symbol	Dimension
1	Stature (a)	Independent	a	$[M^0LT^0]$
2	Shoulder Height (b)	Independent	b	$[M^0LT^0]$
3	Elbow Height (c)	Independent	c	$[M^0LT^0]$
4	Eye Height (d)	Independent	d	$[M^0LT^0]$
5	Finger tip Height (e)	Independent	e	$[M^0LT^0]$
6	Shoulder Breadth (f)	Independent	f	$[M^0LT^0]$
7	Hip Breadth (g)	Independent	g	$[M^0LT^0]$
8	Hand Breadth across thumb (h)	Independent	h	$[M^0LT^0]$
9	Walking Length (W_L)	Independent	W_L	$[M^0LT^0]$
10	Walking Breadth (W_w)	Independent	W_w	$[M^0LT^0]$
11	Anthropometric data (A_1)	Independent	A_1	$[M^0L^0T^0]$
12	Number of Miners (N)	Independent	N	$[M^0L^0T^0]$
13	Age of the Miner (A_m)	Independent	A_m	$[M^0L^0T^0]$
14	Experience in performing work (A_2)	Independent	A_2	$[M^0L^0T^0]$
15	Skills in performing work (A_3)	Independent	A_3	$[M^0L^0T^0]$
16	Posture adopted by Worker (A_4)	Independent	A_4	$[M^0L^0T^0]$
17	Enthusiasm of Performance (A_5)	Independent	A_5	$[M^0L^0T^0]$
18	Habits (A_6)	Independent	A_6	$[M^0L^0T^0]$
19	General health status (A_7)	Independent	A_7	$[M^0L^0T^0]$
20	Diameter of Drill rod (Dr)	Independent	Dr	$[M^0LT^0]$
21	Length of Drill rod (Lr)	Independent	Lr	$[M^0LT^0]$
22	Weight of Drill rod (Wr)	Independent	Wr	$[MLT^{-2}]$
23	Hardness of Drill rod (Hr)	Independent	Hr	$[ML^{-1}T^{-2}]$
24	Diameter of Comp. air Hose (Dc)	Independent	Dc	$[M^0LT^0]$
25	Air Velocity (Ar)	Independent	Ar	$[M^0LT^{-1}]$
26	Length of Comp. air Hose (Lc)	Independent	Lc	$[M^0LT^0]$
27	Weight of Comp. air hose (Wc)	Independent	Wc	$[MLT^{-2}]$
28	Rate of Water flow through hose (Qw)	Independent	Qw	$[M^0L^3T^{-1}]$
29	Weight of Jack hammer (Wj)	Independent	Wj	$[MLT^{-2}]$
30	Illumination (I)	Independent	I	$[M^1L^0T^{-3}]$
31	Speed of Machine (N)	Independent	N	$[M^0L^0T^{-1}]$
32	Penetration rate (R)	Independent	R	$[M^0L^1T^{-1}]$
33	Comp. air Pressure (Pa)	Independent	Pa	$[ML^{-1}T^{-2}]$
34	Ambient temperature (θ)	Independent	θ	$[ML^2T^{-2}]$
35	Relative Humidity (ø)	Independent	ø	$[M^0L^0T^0]$
36	Shear strength of Ore (So)	Independent	So	$[ML^{-1}T^{-2}]$
37	Shear strength of Mica Schist (Ss)	Independent	Ss	$[ML^{-1}T^{-2}]$
40	Density of Ore (Do)	Independent	Do	$[ML^{-3}T^0]$
41	Density of Mica Schist (Ds)	Independent	Ds	$[ML^{-3}T^0]$
42	Time of drilling (Td)	Dependent	Td	$[M^0L^0T^1]$
43	Productivity of drilling (Pd)	Dependent	Pd	$[M^0L^0T^{-1}]$
44	Human energy (He)	Dependent	He	$[ML^2T^{-2}]$

4.3.4 Independent and Dependent π Term

Anthropometric dimensions of the Miner

$$\pi_1 = \left[(N * A_2 * A_4 * A_6 * A_1)/(A_m * A_3 * A_5 * A_7) \right]$$

Where $A1 = \left[a * c * e * g * W_L \right]/\left[b * d * f * h * W_W \right]$

$$\pi_2 = \left[(Dr)^a (Pa)^b (R)^c \right] Lr$$

$$\left[M^0 L^0 T^0 \right] = \left[M^0 L^1 T^0 \right]^a \left[M^1 L^{-1} T^{-2} \right]^b \left[M^0 L^1 T^{-1} \right]^c \left[M^0 L^1 T^0 \right]^l$$

$$M = 0 = 0 + b + 0 + 0, b = 0$$

$$L = 0 = a - b + c + 1,$$

$$T = 0 = 0 - 2b - c + 0, -2b = c, c = 0$$

$$\therefore a - 0 + 0 = -1; a = -1$$

$$\pi_2 = \left[Lr/Dr \right]$$

$$\pi_3 = \left[(Dr)^a (Pa)^b (R)^c \right] Wr$$

$$\left[M^0 L^0 T^0 \right] = \left[M^0 L^1 T^0 \right]^a \left[M^1 L^{-1} T^{-2} \right]^b \left[M^0 L^1 T^{-1} \right]^c \left[M^1 L^1 T^{-2} \right]$$

$$M = 0 = 0 + b + 0 + 1, b = -1$$

$$L = 0 = a - b + c + 1,$$

$$T = 0 = 0 - 2b - c - 2, 2 - c - 2 = 0, c = 0$$

$$\therefore a + 1 + 0 + 1 = 0; a = -2$$

$$\pi_3 = \left[Wr/(Dr^2 * Pa) \right]$$

$$\pi_4 = \left[(Dr)^a (Pa)^b (R)^c \right] Hr$$

$$\left[M^0 L^0 T^0 \right] = \left[M^0 L^1 T^0 \right]^a \left[M^1 L^{-1} T^{-2} \right]^b \left[M^0 L^1 T^{-1} \right]^c \left[M^1 L^{-1} T^{-2} \right]$$

$$M = 0 = 0 + b + 0 + 1, b = -1$$

$$L = 0 = a - b + c + 1,$$

$$T = 0 = 0 - 2b - c - 2, 2 - c - 2 = 0, c = 0$$

$$\therefore a + 1 + 0 - 1 = 0; a = 0$$

$$\pi_4 = \left[Hr/Pa \right]$$

$$\pi_5 = \left[(Dr)^a (Pa)^b (R)^c \right] Dc$$

$$\left[M^0 L^0 T^0 \right] = \left[M^0 L^1 T^0 \right]^a \left[M^1 L^{-1} T^{-2} \right]^b \left[M^0 L^1 T^{-1} \right]^c \left[M^0 L^1 T^0 \right]^1$$

$$M = 0 = 0 + b + 0 + 0, b = 0$$

$$L = 0 = a - b + c + 1,$$

$$T = 0 = 0 - 2b - c + 0, -2b = c, c = 0$$

$$\therefore a - 0 + 0 = -1; a = -1$$

$$\pi_5 = \left[Dc/Dr \right]$$

$$\pi_6 = \left[(Dr)^a (Pa)^b (R)^c \right] Ar$$

$$\left[M^0 L^0 T^0 \right] = \left[M^0 L^1 T^0 \right]^a \left[M^1 L^{-1} T^{-2} \right]^b \left[M^0 L^1 T^{-1} \right]^c \left[M^0 L^1 T^{-1} \right]$$

$$M = 0 = 0 + b + 0 + 0, b = 0$$

$$L = 0 = a - b + c + 1,$$

$$T = 0 = 0 - 2b - c - 1, 0 - c - 1 = 0, c = -1$$

$$\therefore a - 0 - 1 + 1 = 0; a = 0$$

$$\pi_6 = \left[Ar/R \right]$$

$$\pi_7 = \left[(Dr)^a (Pa)^b (R)^c \right] Lc$$

$$\left[M^0 L^0 T^0 \right] = \left[M^0 L^1 T^0 \right]^a \left[M^1 L^{-1} T^{-2} \right]^b \left[M^0 L^1 T^{-1} \right]^c \left[M^0 L^1 T^0 \right]^1$$

$$M = 0 = 0 + b + 0 + 0, b = 0$$

$$L = 0 = a - b + c + 1,$$

$$T = 0 = 0 - 2b - c + 0, -2b = 0, c = 0$$

$$\therefore a - 0 + 0 = -1; a = -1$$

$$\pi_7 = \left[Lc/Dr \right]$$

$$\pi_8 = \left[(Dr)^a (Pa)^b (R)^c \right] Wc$$

$$\left[M^0 L^0 T^0 \right] = \left[M^0 L^1 T^0 \right]^a \left[M^1 L^{-1} T^{-2} \right]^b \left[M^0 L^1 T^{-1} \right]^c \left[M^1 L^1 T^{-2} \right]$$

$$M = 0 = 0 + b + 0 + 1, b = -1$$

$$L = 0 = a - b + c + 1,$$

$$T = 0 = 0 - 2b - c - 2, 2 - c - 2 = 0, c = 0$$

$$\therefore a + 1 + 0 + 1; a = -2$$

$$\pi_8 = \left[Wc / \left(Dr^2 * Pa \right) \right]$$

$$\pi_9 = \left[(Dr)^a (Pa)^b (R)^c \right] Qw$$

$$\left[M^0 L^0 T^0 \right] = \left[M^0 L^1 T^0 \right]^a \left[M^1 L^{-1} T^{-2} \right]^b \left[M^0 L^1 T^{-1} \right]^c \left[M^0 L^3 T^{-1} \right]$$

$$M = 0 = 0 + b + 0 + 0, b = 0$$

$$L = 0 = a - b + c + 3,$$

$$T = 0 = 0 - 2b - c - 1, 0 - c - 1 = 0, c = -1$$

$$\therefore a + 0 - 1 + 3 = 0; a = -2$$

$$\pi_9 = \left[Qw / \left(Dr^2 * R \right) \right]$$

$$\pi_{10} = \left[(Dr)^a (Pa)^b (R)^c \right] Wj$$

$$\left[M^0 L^0 T^0 \right] = \left[M^0 L^1 T^0 \right]^a \left[M^1 L^{-1} T^{-2} \right]^b \left[M^0 L^1 T^{-1} \right]^c \left[M^1 L^1 T^{-2} \right]$$

$$M = 0 = 0 + b + 0 + 1, b = -1$$

$$L = 0 = a - b + c + 1,$$

$$T = 0 = 0 - 2b - c - 2, 2 - c - 2 = 0, c = 0$$

$$\therefore a + 1 + 0 + 1 = 0; a = -2$$

$$\pi_{10} = \left[Wj / \left(Dr^2 * Pa \right) \right]$$

$$\pi_{11} = \left[(Dr)^a (Pa)^b (R)^c \right] I$$

$$\left[M^0 L^0 T^0 \right] = \left[M^0 L^1 T^0 \right]^a \left[M^1 L^{-1} T^{-2} \right]^b \left[M^0 L^1 T^{-1} \right]^c \left[M^1 L^0 T^{-3} \right]$$

$$M = 0 = 0 + b + 0 + 1, b = -1$$

$$L = 0 = a - b + c + 0,$$

$$T = 0 = 0 - 2b - c - 3, 2 - c - 3 = 0, c = -1$$

$$\therefore a + 1 + c + 1 = 0; a + 1 - 1 = 0; a = 0$$

$$\boldsymbol{\pi_{11}} = \left[\mathbf{I}/(\mathbf{Pa} * \mathbf{R}) \right]$$

$$\pi_{12} = \left[(Dr)^a (Pa)^b (R)^c \right] N$$

$$\left[M^0 L^0 T^0 \right] = \left[M^0 L^1 T^0 \right]^a \left[M^1 L^{-1} T^{-2} \right]^b \left[M^0 L^1 T^{-1} \right]^c \left[M^1 L^0 T^{-1} \right]$$

$$M = 0 = 0 + b + 0 + 1, b = 0$$

$$L = 0 = a - b + c + 0,$$

$$T = 0 = 0 - c - 1; c = -1$$

$$\therefore a + 0 + c + 1 = 0; a + 0 - 1 = 0; a = 1$$

$$\boldsymbol{\pi_{12}} = \left[(\mathbf{N} * \mathbf{Dr})/\mathbf{R} \right]$$

$$\boldsymbol{\pi_{13}} = \left[(\theta) \right];$$

$$\boldsymbol{\pi_{14}} = \left[(\phi) \right]$$

$$\pi_{15} = \left[(Dr)^a (Pa)^b (R)^c \right] So$$

$$\left[M^0 L^0 T^0 \right] = \left[M^0 L^1 T^0 \right]^a \left[M^1 L^{-1} T^{-2} \right]^b \left[M^0 L^1 T^{-1} \right]^c \left[M^1 L^{-1} T^{-2} \right]$$

$$M = 0 = 0 + b + 0 + 1, b = -1$$

$$L = 0 = a - b + c + 1,$$

$$T = 0 = 0 - 2b - c - 2, 2 - c - 2 = 0, c = 0$$

$$\therefore a + 1 + 0 - 1 = 0; a = 0$$

$$\boldsymbol{\pi_{15}} = [\mathbf{So}]$$

$$\pi_{16} = \left[So/Pa \right]$$

$$\pi_{17} = \left[(Dr)^a (Pa)^b (R)^c \right] Ss$$

$$\left[M^0 L^0 T^0 \right] = \left[M^0 L^1 T^0 \right]^a \left[M^1 L^{-1} T^{-2} \right]^b \left[M^0 L^1 T^{-1} \right]^c \left[M^1 L^{-1} T^{-2} \right]$$

$$M = 0 = 0 + b + 0 + 1, b = -1$$

$$L = 0 = a - b + c + 1,$$

$$T = 0 = 0 - 2b - c - 2, 2 - c - 2 = 0, c = 0$$

$$\therefore a + 1 + 0 - 1 = 0; a = 0$$

$$\pi_{17} = \left[Ss/Pa \right]$$

$$\pi_{18} = \left[(Dr)^a (Pa)^b (R)^c \right] Do$$

$$\left[M^0 L^0 T^0 \right] = \left[M^0 L^1 T^0 \right]^a \left[M^1 L^{-1} T^{-2} \right]^b \left[M^0 L^1 T^{-1} \right]^c \left[M^0 L^{-3} T^0 \right]$$

$$M = 0 = 0 + b + 0 + 1, b = -1$$

$$L = 0 = a - b + c - 3,$$

$$T = 0 = 0 - 2b - c + 0, 2 - c = 0, c = 2$$

$$\therefore a + 1 + 2 - 3 = 0; a = 0$$

$$\pi_{18} = \left[(Do * R^2)/Pa \right]$$

$$\pi_{19} = \left[(Dr)^a (Pa)^b (R)^c \right] Ds$$

$$\left[M^0 L^0 T^0 \right] = \left[M^0 L^1 T^0 \right]^a \left[M^1 L^{-1} T^{-2} \right]^b \left[M^0 L^1 T^{-1} \right]^c \left[M^0 L^{-3} T^0 \right]$$

$$M = 0 = 0 + b + 0 + 1, b = -1$$

$$L = 0 = a - b + c - 3,$$

$$T = 0 = 0 - 2b - c + 0, 2 - c = 0, c = 2$$

$$\therefore a + 1 + 2 - 3 = 0; a = 0$$

$$\pi_{19} = \left[(Ds * R^2)/Pa \right]$$

$$\pi_{D1} = \left[(Dr)^a (Pa)^b (R)^c \right] Td$$

$$\left[M^0 L^0 T^0 \right] = \left[M^0 L^1 T^0 \right]^a \left[M^1 L^{-1} T^{-2} \right]^b \left[M^0 L^1 T^{-1} \right]^c \left[M^0 L^0 T^1 \right]$$

$$M = 0 = 0 + b + 0 + 0, b = 0$$

$$L = 0 = a - b + c + 0,$$

$$T = 0 = 0 - 2b - c + 1, 0 - c + 1 = 0, c = 1$$

$$\therefore a + 0 + 1 = 0; a = -1$$

$$\pi_{D1} = \left[(Td * R)/Dr \right]$$

$$\pi_{D2} = \left[(Dr)^a (Pa)^b (R)^c \right] Pd$$

$$\left[M^0 L^0 T^0 \right] = \left[M^0 L^1 T^0 \right]^a \left[M^1 L^{-1} T^{-2} \right]^b \left[M^0 L^1 T^{-1} \right]^c \left[M^0 L^0 T^1 \right]$$

$$M = 0 = 0 + b + 0 + 0, b = 0$$

$$L = 0 = a - b + c + 0,$$

$$T = 0 = 0 - 2b - c - 1, 0 - c - 1 = 0, c = -1$$

$$\therefore a + 0 - 1 = 0; a = 1$$

$$\pi_{D2} = \left[(Pd * Dr)/R \right]$$

$$\pi_{D3} = \left[(Dr)^a (Pa)^b (R)^c \right] He$$

$$\left[M^0 L^0 T^0 \right] = \left[M^0 L^1 T^0 \right]^a \left[M^1 L^{-1} T^{-2} \right]^b \left[M^0 L^1 T^{-1} \right]^c \left[M^1 L^2 T^{-2} \right]$$

$$M = 0 = 0 + b + 0 + 1, b = -1$$

$$L = 0 = a - b + c + 2,$$

$$T = 0 = 0 - 2b - c - 2, 2 - c - 2 = 0, c = 0$$

$$\therefore a + 1 + 0 + 2 = 0; a = -3$$

$$\pi_{D3} = \left[He/(Dr^3 * Pa) \right]$$

4.3.5   ESTABLISHMENT OF DIMENSIONLESS GROUP OF π TERMS

These independent variables have been reduced into group of π terms. List of the independent and dependent π terms of the face drilling activity are shown in Tables 4.4 and 4.5:

TABLE 4.4
Independent Dimensionless π Terms

Sr. No.	Independent Dimensionless Ratios	Nature of Basic Physical Quantities
01	$\pi_1 = \left[(N * A_2 * A_4 * A_6 * A_1) / (A_m * A_3 * A_5 * A_7) \right]$ Where $A_1 = [a*c*e*g* W_L]/[b*d*f*h* W_w]$	Anthropometric dimensions of the Miner
02	$\pi_2 = \left[Lr * Dc * Lc/Dr^3 \right]$	Specifications of Drill Rod
03	$\pi_3 = \left[\left\{ (Wr * Wc * Wj)/(Dr^2 * Pa)^3 \right\} * \left\{ (So * Ss * Hr)/(Pa)^3 \right\} * \left\{ (Do * R^2)/Pa \right\} * \left\{ (Ds * R^2)/(Pa) \right\} \right]$	Specifications of Drilling Machine/ process parameters
04	$\pi_4 = \left[\left\{ (N * Dr * Ar)/(R)^2 \right\} * \left\{ (Qw/Dr^2 * R) \right\} \right]$	Speed & Penetration rate of Drill Machine
05	$\pi_5 = [\theta * \phi]$	Ambient temperature and Relative Humidity
06	$\pi_6 = I/[Pa * R]$	Illumination

TABLE 4.5
Dependent Dimensionless π Terms

Sr. No.	Dependent Dimensionless Ratios or π Terms	Nature of Basic Physical Quantities
01	$Z_1 = \left[Td * R/Dr \right]$	Time of drilling
02	$Z_2 = \left[Pd * Dr/R \right]$	Productivity of drilling
03	$Z_3 = \left[He/Dr^3 * Pa \right]$	Human energy

4.3.6 FORMULATION OF FIELD DATA-BASED MODEL

Six independent π terms ($\pi_1, \pi_2, \pi_3, \pi_4, \pi_5, \pi_6$) and three dependent π terms (Z_1, Z_2, Z_3) have been identified for model formulation of field study.

Each dependent π term is a function of the independent π terms,

$$Z_1 = \text{function of} \left(\Pi_1, \Pi_2, \Pi_3, \Pi_4, \Pi_5, \Pi_6 \right)$$

$$Z_2 = \text{function of} \left(\Pi_1, \Pi_2, \Pi_3, \Pi_4, \Pi_5, \Pi_6 \right)$$

$$Z_3 = \text{function of} \left(\Pi_1, \Pi_2, \Pi_3, \Pi_4, \Pi_5, \Pi_6 \right)$$

Where

$Z_1 = (\Pi_{D1})$, First dependent π term $= Td * R/Dr$

$Z_2 = (\Pi_{D2})$, Second dependent π term $= Pd * Dr/R$

$Z_3 = (\Pi_{D3})$, Third dependent π term $= He/Dr^3 * Pa$

The probable exact mathematical form for the dimensional equations of the phenomenon could be relationships assumed to be of exponential form.

$$(Z) = K * \left[(N * A_2 * A_4 * A_6 * A_1)/(A_m * A_3 * A_5 * A_7) \right]^a , \left[(Lr * DC * Lc)/(Dr \right.$$

$$\left[\left\{ (Wr * Wc * Wj)/(Dr^2 * Pa)^3 \right\} * \left\{ (So * Ss * Hr)/(Pa)^3 \right\} \right.$$

$$* \left\{ (Do * R^2)/Pa \right\} * \left\{ (Ds * R^2)/(Pa) \right\} \Big]^c ,$$

$$\left[\left\{ (N * Dr * Ar)/(R) \right\} * \left\{ (Qw/Dr^2 * R) \right\} \right]^d , \left[(\theta) * (\phi) \right]^e , \left[I/(Pa * R) \right]^f \qquad (4.1)$$

4.3.7 MODEL FORMULATION BY IDENTIFYING THE CURVE FITTING CONSTANT & VARIOUS INDICES OF π TERMS

The multiple regression analysis helps to identify the indices of the different π terms in the model aimed at, by considering six independent π terms and one dependent π term. Let the model aimed at being of the form,

$$(Z_1) = K_1 * \left[(\pi_1)^{a1} * (\pi_2)^{b1} * (\pi_3)^{c1} * (\pi_4)^{d1} * (\pi_5)^{e1} * (\pi_6)^{f1} \right] \qquad (4.2)$$

$$(Z_2) = K_2 * \left[(\pi_1)^{a2} * (\pi_2)^{b2} * (\pi_3)^{c2} * (\pi_4)^{d2} * (\pi_5)^{e2} * (\pi_6)^{f2} \right] \qquad (4.3)$$

$$(Z_3) = K_3 * \left[(\pi_1)^{a3} * (\pi_2)^{b3} * (\pi_3)^{c3} * (\pi_4)^{d3} * (\pi_5)^{e3} * (\pi_6)^{f3} \right] \qquad (4.4)$$

To find the values of a_1, b_1, c_1, d_1, e_1, and f_1, Equation 3.2 is presented as follows:

$$\sum Z_1 = nK_1 + a_1 * \sum A + b_1 * \sum B + c_1 * \sum C + d_1 * \sum D + e_1 * \sum E + f_1 * \sum F$$

$$\sum Z_1 * A = K_1 * \sum A + a_1 * \sum A * A + b_1 * \sum B * A + c_1 * \sum C * A$$

$$+ d_1 * \sum D * A + e_1 * \sum E * A + f_1 * \sum F * A$$

$$\sum Z_1 * B = K_1 * \sum B + a_1 * \sum A * B + b_1 * \sum B * B + c_1 * \sum C * B$$

$$+ d_1 * \sum D * B + e_1 * \sum E * B + f_1 * \sum F * B$$

$$\sum Z_1 * C = K_1 * \sum C + a_1 * \sum A * C + b_1 * \sum B * C + c_1 * \sum C * C$$

$$+ d_1 * \sum D * C + e_1 * \sum E * C + f_1 * \sum F * C$$

$$\sum Z_1 * D = K_1 * \sum D + a_1 * \sum A * D + b_1 * \sum B * D + c_1 * \sum C * D$$

$$+ d_1 * \sum D * D + e_1 * \sum E * D + f_1 * \sum F * D$$

$$\sum Z_1 * E = K_1 * \sum E + a_1 * \sum A * E + b_1 * \sum B * E + c_1 * \sum C * E$$

$$+ d_1 * \sum D * E + e_1 * \sum E * E + f_1 * \sum F * E$$

$$\sum Z_1 * F = K_1 * \sum F + a_1 * \sum A * F + b_1 * \sum B * F + c_1 * \sum C * F$$

$$+ d_1 * \sum D * F + e_1 * \sum E * F + f_1 * \sum F * F$$

In the above set of equations, the values of K_1, a_1, b_1, c_1, d_1, e_1, and f_1 are substituted to compute the values of the unknowns. After substituting these values in the equations, one will get a set of seven equations, which are to be solved simultaneously to get the values of K_1, a_1, b_1, c_1, d_1, e_1, and f_1. The abovementioned equations can be transfer used in the matrix form and subsequently values of K_1, a_1, b_1, c_1, d_1, e_1, and f_1 can be obtained by adopting matrix analysis.

$$X_1 = \text{inv}(W) \times P_1$$

W = 7×7 matrix of the multipliers of K_1, a_1, b_1, c_1, d_1, e_1, and f_1
$P_1 = 7 \times 1$ matrix of the terms on L H S and
$X_1 = 7 \times 1$ matrix of solutions of values of K_1, a_1, b_1, c_1, d_1, e_1, and f_1

Then, the matrix obtained is given by,
  Matrix

$$
Z_1 \times
\begin{bmatrix} 1 \\ A \\ B \\ C \\ D \\ E \\ F \end{bmatrix}
=
\begin{bmatrix}
n & A & B & C & D & E & F \\
A & A^2 & BA & CA & DA & EA & FA \\
B & AB & B^2 & CB & DB & EB & FB \\
C & AC & BC & C^2 & DC & EC & FC \\
D & AD & BD & CD & D^2 & ED & FD \\
E & AE & BE & CE & DE & E^2 & FE \\
F & AF & BF & CF & DF & EF & F^2
\end{bmatrix}
\times
\begin{bmatrix} K_1 \\ a_1 \\ b_1 \\ c_1 \\ d_1 \\ e_1 \\ f_1 \end{bmatrix}
$$

X_1 matrix with K_1 and indices a_1, b_1, c_1, d_1, e_1, and f_1 evaluated:

In the abovementioned equations is the number of sets of readings, A, B, C, D, E, and F represent the independent π terms π_1, π_2, π_3, π_4, π_5, and π_6 while, Z represents dependent π term.

4.4   BASIS FOR ARRIVING AT NUMBER OF OBSERVATIONS

The number of observations taken is 30 based on the probability concept of Degree of Uncertainty. The formula for calculating the number of readings is

$$\sqrt{N} = \left[\left\{x/Zc\right\} - \mu\right]\sigma$$

Where

x = Mean,

μ = Median

σ = Standard Deviation

N = Number of Readings

For N ≥ 30, Zc = 2.58 for Certainty (Confidence level 99%)

Zc = 1.96 for Certainty (Confidence level 95%)

Zc = 1.645 for Certainty (Confidence level 90%)

Selecting Zc = 2.58 for Certainty with the Confidence level 99%, satisfied the no. of readings

$$N = \left[\left\{x/Tc\right\} - \mu\right]^2 \sigma$$

For N < 30, Ţc = 2.48 for Certainty (Confidence level 99%)

Ţc = 1.71 for Certainty (Confidence level 95%)

Ţc = 1.32 for Certainty (Confidence level 90%)

Selecting Ţc = 2.48 for Certainty with the Confidence level 99%, satisfied the number of readings.

5 Procedure of Collecting Field Data: Causes, Extraneous Variables, and Effects

Pramod Belkhode
Laxminarayan Institute of Technology

J. P. Modak
Visvesvaraya National Institute of Technology and
JD College of Engineering and Management

V. Vidyasagar
Power Systems Training Institute

Sagar Shelare
Priyadarshini College of Engineering

CONTENTS

5.1 AN APPROACH TO FORMULATE THE MATHEMATICAL MODEL

The procedure of data collection, instrumentation used, and the data collected from field have been presented in this chapter. The identified independent variables have been reduced into groups of dimensionless π terms. The developed models show the relationship among the dimensionless π terms of the phenomenon.

5.2 INSTRUMENTATION AND DATA COLLECTION

5.2.1 Instrumentation for Face Drilling

The face drilling activity (hereafter nominated as phenomena) phenomena are influenced by the following variables (Table 5.1). The instrumentation used for these variables for measurement is associated as follows:

TABLE 5.1
Independent Variables and Instrumentation-Face Drilling

Sr. No.	Description	Type of Variables	Symbol	Dimension	Dimension Measurement
1	Stature (a)	Independent	a	$[M^0LT^0]$	Measuring tape Stadiometer
2	Shoulder height (b)	Independent	b	$[M^0LT^0]$	Measuring tape
3	Elbow height (c)	Independent	c	$[M^0LT^0]$	Measuring tape
4	Eye height (d)	Independent	d	$[M^0LT^0]$	Measuring tape
5	Fingertip height (e)	Independent	e	$[M^0LT^0]$	Measuring tape
6	Shoulder breadth (f)	Independent	f	$[M^0LT^0]$	Measuring tape
7	Hip breadth (g)	Independent	g	$[M^0LT^0]$	Measuring tape
8	Hand breadth across thumb (h)	Independent	h	$[M^0LT^0]$	Transparent scale (30 cm)
9	Walking length (W_L)	Independent	W_L	$[M^0LT^0]$	Measuring tape
10	Walking breadth (W_W)	Independent	W_W	$[M^0LT^0]$	Measuring tape
11	Anthropometric data (A_1)	Independent	A_1	$[M^0L^0T^0]$	Measured from formula
12	Number of miners (N)	Independent	N	$[M^0L^0T^0]$	Team will have 5 miners
13	Age of the miner (A_m)	Independent	A_m	$[M^0L^0T^0]$	Questionnaire
14	Experience in performing work (A_2)	Independent	A_2	$[M^0L^0T^0]$	Questionnaire
15	Skills in performing work (A_3)	Independent	A_3	$[M^0L^0T^0]$	Observation of supervisor
16	Posture adopted by Worker (A_4)	Independent	A_4	$[M^0L^0T^0]$	Observation
17	Enthusiasm in performing the activity (A_5)	Independent	A_5	$[M^0L^0T^0]$	Questionnaire
18	Habits (A_6)	Independent	A_6	$[M^0L^0T^0]$	Questionnaire
19	General health status (A_7)	Independent	A_7	$[M^0L^0T^0]$	Questionnaire
20	Diameter of drill rod (Dr)	Independent	Dr	$[M^0LT^0]$	Vernier Calipers
21	Length of drill rod (Lr)	Independent	Lr	$[M^0LT^0]$	Measuring Tape
22	Weight of drill rod (Wr)	Independent	Wr	$[MLT^{-2}]$	Weighing Balance (Digital)
23	Hardness of drill rod (Hr)	Independent	Hr	$[ML^{-1}T^{-2}]$	Vernier Calipers
24	Diameter of compressed air Hose (Dc)	Independent	Dc	$[M^0LT^0]$	Vernier Calipers

(Continued)

TABLE 5.1 (Continued)
Independent Variables and Instrumentation-Face Drilling

Sr. No.	Description	Type of Variables	Symbol	Dimension	Dimension Measurement
25	Air velocity (Ar)	Independent	Ar	$[M^0LT^{-1}]$	Ultrasonic Anemometer
26	Length of compressed air Hose (Lc)	Independent	Lc	$[M^0LT^0]$	Measuring Tape
27	Weight of compressed air hose (Wc)	Independent	Wc	$[MLT^{-2}]$	Specific Length Cut and weighed
28	Rate of Water flowthrough hose (Qw)	Independent	Qw	$[M^0L^3T^{-1}]$	Flow meter
29	Weight of Jack hammer (Wj)	Independent	Wj	$[MLT^{-2}]$	Weighed on Digital Balance
30	Illumination (I)	Independent	I	$[M^1L^0T^{-3}]$	Lux Meter
31	Speed of machine (N)	Independent	N	$[M^0L^0T^{-1}]$	Tachometer
32	Penetration rate (R)	Independent	R	$[M^0L^1T^{-1}]$	Measured
33	Compressed air pressure (Pa)	Independent	Pa	$[ML^{-1}T^{-2}]$	Readings from Compressor Room
34	Ambient temperature (θ)	Independent	θ	$[ML^2T^{-2}]$	Temperature measuring Meter
35	Relative humidity (ø)	Independent	ø	$[M^0L^0T^0]$	Hygrometer
36	Shear strength of ore (So)	Independent	So	$[ML^{-1}T^{-2}]$	Measured in Laboratory
37	Shear strength of Mica Schist (Ss)	Independent	Ss	$[ML^{-1}T^{-2}]$	Measured in Laboratory
40	Density of ore (Do)	Independent	Do	$[ML^{-3}T^0]$	Measured in Laboratory
41	Density of Mica Schist (Ds)	Independent	Ds	$[ML^{-3}T^0]$	Measured in Laboratory
42	Time of drilling (Td)	Dependent	Td	$[M^0L^0T^1]$	Calculated
43	Productivity of drilling (Pd)	Dependent	Pd	$[M^0L^0T^{-1}]$	Calculated
44	Human energy (He)	Dependent	He	$[ML^2T^{-2}]$	Calculated from heartbeats/min.

The field data has been collected from underground mines. The readings have been collected from 33.0 m, 66.0 m and 100 m levels at six workstations with a team of five miners at each location at different timings.

A total of 30 miners have been enquired through Questionnaire for the data such as Age of the miner (Am), Experience in performing work (A_2), general health status (A_7) and habits (A_6) are rated on a 1–10 scale. Three supervisors have also been enquired through questionnaire for collecting data such as skills in performing work (A_3), enthusiasm in performing the activity (A_5) of the individual miner to rate them on 1–10 scale. Postures have been arrived at by observing the activity at the workstations and are numerically rated as 1 and 2 for standing erect and bending, respectively.

Weight of the individual miner has been measured using a digital weighing balance. The measurements such as stature (a), shoulder height (b), elbow height (c), eye height (d), fingertip height (e), shoulder breadth (f), hip breadth (g), hand breadth across thumb (h), walking length (W_L), and walking breadth (W_W) have been measured using Harpenden Stadiometer, Holtain Bicondylar Calipers and measuring tape for body.

Measuring very low air velocities in underground metallic ore mines: the technique of mounting the lightweight Airflow UA-6 sensor probe to the top of an extensible/graduated rod, bringing a nonstandard 35-ft-long connecting cable down the rod and reading the air velocity at a convenient height above the mine floor was developed.

Parameters such as length of drill rod (Lr), length of compressed air hose (Lc) have been measured with a 10 m measuring tape. The diameter of drill rod (Dr) and diameter of compressed air hose (Dc) have been measured using Vernier Callipers. Then, the values of π_2 have been arrived at by calculation.

Parameters such as weight of drill rod (Wr), weight of Jack hammer (Wj), and specific size of compressed air hose have been cut and weighed on digital weighing balance to arrive at the weight of compressed air hose (Wc) [18]. The properties such as shear strength of Ore (So), shear strength of Mica Schist (Ss), density of ore (Do), and density of Mica Schist (Ds) have been collected from Laboratory. The hardness of drill rod (Hr) has been collected from specifications provided by the manufacturer. The compressed air pressure (Pa) has been collected as a real-time parameter from compressor Instrument panel. Ambient air velocity (very low) in underground mines has been measured with the Airflow UA-6 sensor probe. The values of π_3 have been arrived at by calculation.

Parameter such as speed of the drilling machine (N) has been measured using Tachometer. Time for drilling has been measured using stopwatch on smart phone. The penetration rate (R) has been measured using Vernier Calipers. Then, the values of π_4 have been arrived at by calculation.

The ambient temperature (Dry bulb) and relative humidity have been measured with the dry bulb temperature and relative humidity measuring digital meter [24,25]. The illumination (I) at the workstation has been measured using light meter. These readings have been used for calculating π_5, π_6, and π_7 terms.

The human energy consumed has been arrived at by measuring the heart rate/min with a polar heart rate monitor.

The heart rate is measured during the work to know the human energy consumption. The "polar" make heart rate monitor is used for measuring the heartbeats. The accuracy of this equipment is ±1% or ±1 bpm, whichever is larger under steady-state condition. The total working time is 9 hours 59 minutes. For less than 1 hour, the display indicates mm:ss; for more than 1 hour, the display indicates hh:mm. The minimum duration of recorded time is 1 minute. The transmitter is placed on the chest just below the chest muscles with elastic strap. The wrist receiver can be attached to wrist or kept at a suitable position within 1-m range from the transmitter. The human energy can be estimated by many methods. However, the heart rate measurement is the simplest and most suitable in the context of this research for estimation of the human energy [26, 27]. It is calculated by using the scheme of relationship between energy expenditure and heart rate.

Procedure for measuring the heartbeats

The working heart rate (WHR) was measured by placing a stethoscope at the apex of the heart; time was measured using a stopwatch for 10 beats. It was then expressed in beats per minute. WHR was measured at every 4th minute of the interval throughout the working phase. The response variable π terms have been calculated.

5.2.2 Data Collection from the Field for Face Drilling

The field data has been collected from underground mines. The readings have been collected from 33.0, 66.0, m and 100 m depths level at six workstations with a team of five miners at each location are shown in Tables 5.2 to 5.9.

$$HE = 0.6309*(HR\ Max - HR\ Start) + 0.1988*Weight + 0.2017*Age$$

$$Sample\ Calculation: HE = 0.6309*(HR\ Max - HR\ Start)$$

$$+ 0.1988*Weight + 0.2017*Age$$

$$Human\ Energy(KJ) = 0.6309*(240 - 176) + 0.1988*65 + 0.2017*(47)$$

$$= 40.3776 + 12.922 + 9.4799$$

$$= 62.7795\ KJ\ for\ 5\ min.\ or\ 75.335\ for\ 360s\ or\ 6\,min.$$

TABLE 5.2
Anthropometric Data of the Miners

S. No.	Miner No.	Age	Weight	Stature (a) mm	Shoulder Height (b) mm	Elbow Height (c) mm	Eye Height (d) mm	Finger Tip Height (e) mm	Shoulder Breadth (f) mm	Hip Breadth (g) mm	Hand Breadth across Thumb (h)	Walking Length (Wl) mm	Walking Breadth (Ww) mm	Experience Years	Skill	Posture	Enthusiasm	Habits	Health	$A_1 = (a * c * e * g * Wl)/(b * d * f * h * Ww)$
1	1	47	65	1750	1425	1080	1665	660	470	370	123	590	430	20	9	1	5	6	7	4.834773
2	2	45	58	1730	1405	1085	1650	640	460	365	120	585	420	18	8	1	6	7	8	4.968332
3	3	36	55	1770	1440	1090	1620	660	475	375	124	600	435	21	9	1	7	6	7	4.496128
4	4	52	66	1740	1420	1060	1690	650	480	470	123	580	430	25	9	2	4	6	6	5.701565
5	5	42	62	1710	1400	1020	1660	635	470	460	119	570	390	15	7	2	7	8	8	6.095939
6	1	40	82	1850	1549	1030	1753	686	470	406	127	620	400	13	6	2	7	8	8	5.311571
7	2	45	60	1650	1240	890	1475	735	425	355	115	610	385	18	8	2	6	7	7	5.902393
8	3	53	61	1680	1260	950	1490	630	430	340	117	615	390	26	9	1	4.5	6	6	5.707576
9	4	50	60	1700	1270	1020	1520	640	457	355	120	617	410	23	9	1	5	6	6	5.600335
10	5	47	65	1750	1425	1080	1665	660	470	370	120	590	430	20	9	1	6	7	5	4.910074
11	1	48	61	1640	1325	1006	1527	602	420	320	116	585	430	21	9	1	6	6	7	4.386453
12	2	50	65	1750	1427	1080	1638	660	462	360	125	590	435	23	9	1	4	6	6	4.623317
13	3	47	65	1750	1427	1080	1638	660	462	360	125	590	435	20	9	1	5	7	7	4.559343
14	4	45	61	1640	1325	1006	1527	602	420	320	116	585	430	18	8	2	7	7	5	4.386453
15	5	49	65	1750	1427	1080	1638	660	462	360	125	590	435	21	9	2	6	6	7	4.496128

(Continued)

TABLE 5.2 (Continued)
Anthropometric Data of the Miners

S. No.	Miner No.	Age	Weight	Stature (a) mm	Shoulder Height (b) mm	Elbow Height (c) m	Eye Height (d) mm	Finger Tip Height (e) mm	Shoulder Breadth (f) mm	Hip Breadth (g) mm	Hand Breadth across Thumb (h)	Walking Length (WI) mm	Walking Breadth (Ww) mm	Experience Years	Skill	Posture	Enthusiasm	Habits	Health	$A_1 = (a*c*e*g*WI)/(b*d*f*h*Ww)$
16	1	47	61	1640	1325	1006	1527	602	420	320	116	585	430	20	9	2	6	7	5	4.386453
17	2	46	65	1750	1425	1080	1665	660	470	370	120	590	430	19	8	2	6	7	7	4.842442
18	3	52	61	1680	1260	950	1490	630	430	340	117	590	390	25	9	1	4	6	6	5.341928
19	4	46	60	1650	1240	890	1475	735	425	355	115	590	385	19	8	1	6	7	7	5.768039
20	5	50	60	1700	1270	1020	1520	640	457	355	120	617	410	23	9	1	5	6	6	5.576006
21	1	41	66	1710	1400	1020	1660	635	470	460	119	570	390	14	6	1	7	8	8	6.007937
22	2	51	71	1740	1420	1060	1690	650	480	470	123	580	430	21	9	1	5	6	7	4.645237
23	3	37	55	1770	1440	1090	1620	660	475	375	124	600	435	18	8	1	6	7	5	4.509559
24	4	44	58	1730	1405	1085	1650	640	460	365	120	585	420	17	7	2	7	7	8	4.969135
25	5	46	65	1750	1425	1080	1665	660	470	370	123	590	430	21	9	2	4.5	6	7	4.509431
26	1	50	60	1700	1270	1020	1520	640	457	355	120	617	410	23	9	2	4.5	8	6	5.594779
27	2	42	58	1730	1405	1090	1650	640	460	365	120	585	420	15	7	2	7	8	8	5.08052
28	3	35	69	1770	1440	1090	1620	660	475	375	124	600	435	21	9	1	5	6	5	4.436832
29	4	49	66	1740	1420	1060	1690	650	480	470	123	580	430	22	9	1	5	6	7	5.798709
30	5	39	58	1710	1400	1020	1660	635	470	460	119	570	390	12	6	1	7	8	8	6.01217

TABLE 5.3

Data Related to Dimensions of Drill Rod and Compressed Air Hose

S. No.	Length of Drill Rod (Lr) m	Diameter of Compressed Air Hose (Dc) m	Length of Compressed Air Hose (Lc) m	Diameter of Drill Rod (Dr) m
1	0.800	0.019	5.0	0.034
2	0.800	0.019	7.0	0.034
3	0.800	0.019	8.0	0.034
4	0.800	0.019	9.0	0.034
5	0.800	0.019	10.0	0.034
6	1.200	0.019	5.0	0.033
7	1.200	0.019	7.0	0.033
8	1.200	0.019	8.0	0.033
9	1.200	0.019	9.0	0.033
10	1.200	0.019	10.0	0.033
11	1.600	0.019	5.0	0.033
12	1.600	0.019	7.0	0.033
13	1.600	0.019	8.0	0.033
14	1.600	0.019	9.0	0.033
15	1.600	0.019	10.0	0.033
16	0.800	0.019	6.0	0.034
17	0.800	0.019	8.0	0.034
18	0.800	0.019	9.0	0.034
19	0.800	0.019	10.0	0.034
20	0.800	0.019	12.0	0.034
21	1.200	0.019	6.0	0.033
22	1.200	0.019	8.0	0.033
23	1.200	0.019	9.0	0.033
24	1.200	0.019	10.0	0.033
25	1.200	0.019	12.0	0.033
26	1.600	0.019	6.0	0.033
27	1.600	0.019	8.0	0.033
28	1.600	0.019	9.0	0.033
29	1.600	0.019	10.0	0.033
30	1.600	0.019	12.0	0.033

TABLE 5.4
Data Related to Drilling Machine/Process Parameters

S. No.	Weight of Drill Rod (Wr) Kgf	Weight of Compressed Air Hose (Wc) Kgf	Weight of Jack Hammer (Wj) Kgf	Diameter of Drill Rod (Dr) m	Shear Strength of Ore (So) (N/m²)	Shear Strength of Mica Schist (Ss) (N/m²)	Density of Ore (Do) kg/m³	Ambient Air Velocity (Ar) m/s	Density of Mica Schist (Ds) kg/m³	Compressed Air Pressure (Pa)	Hardness of Drill Rod (Hr)	Rate of Water Flow Through Hose (Qw) m³/s
1	2.9	0.9	25	0.034	5,560,000	4,008,000	3500	0.8	2300	6.0	56	1.5714E-05
2	2.9	1.08	25	0.034	5,560,000	4,008,000	3500	0.9	2300	5.9	56	1.5714E-05
3	2.9	1.2	25	0.034	5,560,000	4,008,000	3500	1	2300	5.8	56	1.5714E-05
4	2.9	1.32	25	0.034	5,560,000	4,008,000	3500	1.1	2300	6.1	56	1.5714E-05
5	2.9	1.5	25	0.034	5,560,000	4,008,000	3500	1.2	2300	6.0	56	1.5714E-05
6	4.1	0.9	25	0.033	5,560,000	4,008,000	3500	0.8	2300	6.0	56	1.5714E-05
7	4.1	1.08	25	0.033	5,560,000	4,008,000	3500	0.9	2300	5.9	56	1.5714E-05
8	4.1	1.2	25	0.033	5,560,000	4,008,000	3500	1	2300	5.8	56	1.5714E-05
9	4.1	1.32	25	0.033	5,560,000	4,008,000	3500	1.1	2300	6.1	56	1.5714E-05
10	4.1	1.5	25	0.033	5,560,000	4,008,000	3500	1.2	2300	6.0	56	1.5714E-05
11	5.2	0.9	25	0.033	5,560,000	4,008,000	3500	0.8	2300	6.0	56	1.5714E-05
12	5.2	1.08	25	0.033	5,560,000	4,008,000	3500	0.9	2300	5.9	56	1.5714E-05
13	5.2	1.2	25	0.033	5,560,000	4,008,000	3500	1	2300	5.8	56	1.5714E-05
14	5.2	1.32	25	0.033	5,560,000	4,008,000	3500	1.1	2300	6.1	56	1.5714E-05
15	5.2	1.5	25	0.033	5,560,000	4,008,000	3500	1.2	2300	6.0	56	1.5714E-05
16	2.9	0.9	25	0.034	5,560,000	4,008,000	3500	0.8	2300	6.0	56	1.5714E-05
17	2.9	1.08	25	0.034	5,560,000	4,008,000	3500	0.9	2300	5.9	56	1.5714E-05
18	2.9	1.2	25	0.034	5,560,000	4,008,000	3500	1	2300	5.8	56	1.5714E-05

(Continued)

TABLE 5.4 (*Continued*)
Data Related to Drilling Machine/Process Parameters

S. No.	Weight of Drill Rod (Wr) Kgf	Weight of Compressed Air Hose (Wc) Kgf	Weight of Jack Hammer (Wj) Kgf	Diameter of Drill Rod (Dr) m	Shear Strength of Ore (So) (N/m²)	Shear Strength of Mica Schist (Ss) (N/m²)	Density of Ore (Do) kg/m³	Ambient Air Velocity (Ar) m/s	Density of Mica Schist (Ds) kg/m³	Compressed Air Pressure (Pa)	Hardness of Drill Rod (Hr)	Rate of Water Flow Through Hose (Qw) m³/s
19	2.9	1.32	25	0.034	5,560,000	4,008,000	3500	1.1	2300	6.1	56	1.5714E−05
20	2.9	1.5	25	0.034	5,560,000	4,008,000	3500	1.2	2300	6.0	56	1.5714E−05
21	4.1	0.9	25	0.033	5,560,000	4,008,000	3500	0.8	2300	6.0	56	1.5714E−05
22	4.1	1.08	25	0.033	5,560,000	4,008,000	3500	0.9	2300	5.9	56	1.5714E−05
23	4.1	1.2	25	0.033	5,560,000	4,008,000	3500	1	2300	5.8	56	1.5714E−05
24	4.1	1.32	25	0.033	5,560,000	4,008,000	3500	1.1	2300	6.1	56	1.5714E−05
25	4.1	1.5	25	0.033	5,560,000	4,008,000	3500	1.2	2300	6.0	56	1.5714E−05
26	5.2	0.9	25	0.033	5,560,000	4,008,000	3500	0.8	2300	6.0	56	1.5714E−05
27	5.2	1.08	25	0.033	5,560,000	4,008,000	3500	0.9	2300	5.9	56	1.5714E−05
28	5.2	1.2	25	0.033	5,560,000	4,008,000	3500	1	2300	5.8	56	1.5714E−05
29	5.2	1.32	25	0.033	5,560,000	4,008,000	3500	1.1	2300	6.1	56	1.5714E−05
30	5.2	1.5	25	0.033	5,560,000	4,008,000	3500	1.2	2300	6.0	56	1.5714E−05

TABLE 5.5
Data Related to Speed, Penetration Rate, Time and Ambient Air Velocity

S. No.	Diameter of Drill Rod (Dr) m	Speed of Drill M/c (N)	Penetration Rate (R) m/s	Time (t) S	Air Velocity (Ar) m/s
1	0.034	60	0.00166	360	0.8
2	0.034	49	0.0016	375	0.9
3	0.034	46	0.00154	390	1
4	0.034	64	0.00174	345	1.1
5	0.034	66	0.00164	365	1.2
6	0.033	60	0.00166	600	0.8
7	0.033	56	0.00163	615	0.9
8	0.033	69	0.00159	630	1
9	0.033	62	0.00171	585	1.1
10	0.033	76	0.00169	590	1.2
11	0.033	58	0.00166	840	0.8
12	0.033	60	0.00164	855	0.9
13	0.033	55	0.00161	870	1
14	0.033	62	0.0017	825	1.1
15	0.033	66	0.00167	835	1.2
16	0.034	60	0.00166	360	0.8
17	0.034	49	0.0016	375	0.9
18	0.034	46	0.00154	390	1
19	0.034	64	0.00174	345	1.1
20	0.034	66	0.00164	365	1.2
21	0.033	60	0.00166	600	0.8
22	0.033	56	0.00163	615	0.9
23	0.033	69	0.00159	630	1
24	0.033	62	0.00171	585	1.1
25	0.033	76	0.00169	590	1.2
26	0.033	58	0.00166	840	0.8
27	0.033	60	0.00164	855	0.9
28	0.033	55	0.00161	870	1
29	0.033	62	0.0017	825	1.1
30	0.033	66	0.00167	835	1.2

TABLE 5.6

Data Related to Ambient Temperature, Relative Humidity, and Illumination

S. No.	Ambient Temperature (°C)	Relative Humidity	Illumination (I)
1	27	90	1615
2	27	90	1615
3	27	90	1615
4	27	91	1615
5	27	91	1615
6	28	91	1615
7	28	91	1615
8	28	92	1615
9	28	92	1615
10	28	92	1615
11	29	93	1615
12	29	93	1615
13	29	93	1615
14	29	89	1615
15	29	89	1615
16	27	89	1615
17	27	88	1615
18	27	88	1615
19	27	88	1615
20	27	87	1615
21	28	87	1615
22	28	87	1615
23	28	86	1615
24	28	86	1615
25	28	86	1615
26	29	85	1615
27	29	85	1615
28	29	85	1615
29	29	90	1615
30	29	91	1615

TABLE 5.7

Data Related to Dependent π Terms Time, Productivity and Human Energy Input for Face Drilling

S. No.	Time of Drilling (s)	Penetration Rate (R) (m/s)	Diameter of Drill Rod (m)	Productivity (Pd)	Human Energy KJ (He)
1	360	0.00166	0.034	0.09027	75.335
2	375	0.0016	0.034	0.09444	136.1663
3	390	0.00154	0.034	0.09722	113.8725
4	345	0.00174	0.034	0.10069	72.1338
5	365	0.00164	0.034	0.10416	128.1612
6	600	0.00166	0.033	0.09027	186.2754
7	615	0.00163	0.033	0.09444	120.66
8	630	0.00159	0.033	0.09722	180.4045
9	585	0.00171	0.033	0.10069	175.7925
10	590	0.00169	0.033	0.10416	210.3212
11	840	0.00166	0.033	0.09027	144.9224
12	855	0.00164	0.033	0.09444	173.4538
13	870	0.00161	0.033	0.09722	247.9265
14	825	0.0017	0.033	0.10069	245.685
15	835	0.00167	0.033	0.10416	298.7797
16	360	0.00166	0.034	0.09027	74.3812
17	375	0.0016	0.034	0.09444	138.1575
18	390	0.00154	0.034	0.09722	117.0185
19	345	0.00174	0.034	0.10069	69.3703
20	365	0.00164	0.034	0.10416	129.64
21	600	0.00166	0.033	0.09027	180.3172
22	615	0.00163	0.033	0.09444	127.6238
23	630	0.00159	0.033	0.09722	171.1225
24	585	0.00171	0.033	0.10069	172.6577
25	590	0.00169	0.033	0.10416	209.924
26	840	0.00166	0.033	0.09027	276.587
27	855	0.00164	0.033	0.09444	164.889
28	870	0.00161	0.033	0.09722	243.2134
29	825	0.0017	0.033	0.10069	250.6386
30	835	0.00167	0.033	0.10416	289.2941

TABLE 5.8
Data Related to the Heartbeat Measurement and HE Calculation

Sub	Age (Yrs)	Weight(kgs)	Avg HR	Start	Max.	Human Energy Formula KJ (Total time)	Time of drilling (s)
1	47	65	233	176	240	75.335	360
2	45	58	220	100	240	136.1663	375
3	36	55	195	129	239	113.8725	390
4	52	66	215	174	236	72.1338	345
5	42	62	200	101	235	128.1612	365
6	40	82	185	125	234	186.2754	600
7	45	60	219	178	238	120.66	615
8	53	61	190	135	235	180.4045	630
9	50	60	179	127	235	175.7925	585
10	47	65	200	101	235	210.3212	590
11	48	61	185	125	234	144.9224	840
12	50	65	219	178	238	173.4538	855
13	47	65	190	135	235	247.9265	870
14	45	61	179	127	235	245.685	825
15	49	65	190	101	235	298.7797	835
16	47	61	233	176	240	74.3812	360
17	46	65	220	100	240	138.1575	375
18	52	61	195	129	239	117.0185	390
19	46	60	215	174	236	69.3703	345
20	50	60	200	101	235	129.64	365
21	41	66	185	125	234	180.3172	600
22	51	71	219	178	238	127.6238	615
23	37	55	190	135	235	171.1225	630
24	44	58	179	127	235	172.6577	585
25	46	65	200	101	235	209.924	590
26	50	60	185	125	234	276.587	840
27	42	58	219	178	238	164.889	855
28	35	69	190	135	235	243.2134	870
29	49	66	179	127	235	250.6386	825
30	39	58	190	101	235	289.2941	835

TABLE 5.9

Data Related to Dependent π Term Human Energy for Face Drilling

S. No.	Human Energy KJ (He)	Diameter of Drill Rod (Dr) m	Compressed Air Pressure (Pa)
1	75.335	0.034	6
2	136.1663	0.034	5.9
3	113.8725	0.034	5.8
4	72.1338	0.034	6.1
5	128.1612	0.034	6
6	186.2754	0.033	6
7	120.66	0.033	5.9
8	180.4045	0.033	5.8
9	175.7925	0.033	6.1
10	210.3212	0.033	6
11	144.9224	0.033	6
12	173.4538	0.033	5.9
13	247.9265	0.033	5.8
14	245.685	0.033	6.1
15	298.7797	0.033	6
16	74.3812	0.034	6
17	138.1575	0.034	5.9
18	117.0185	0.034	5.8
19	69.3703	0.034	6.1
20	129.64	0.034	6
21	180.3172	0.033	6
22	127.6238	0.033	5.9
23	171.1225	0.033	5.8
24	172.6577	0.033	6.1
25	209.924	0.033	6
26	276.587	0.033	6
27	164.889	0.033	5.9
28	243.2134	0.033	5.8
29	250.6386	0.033	6.1
30	289.2941	0.033	6

6 Mathematical Modeling of Operations

Pramod Belkhode
Laxminarayan Institute of Technology

J. P. Modak
Visvesvaraya National Institute of Technology and
JD College of Engineering and Management

V. Vidyasagar
Power Systems Training Institute

P. B. Maheshwary
JD College of Engineering and Management

CONTENTS

6.1 FORMULATION OF FIELD DATA-BASED MODEL FOR FACE DRILLING

Six independent π terms (π_1, π_2, π_3, π_4, π_5, and π_6) and three dependent π terms (Z_1, Z_2, and Z_3) have been identified for field study-based model formulation.

Each dependent π term is a function of the available independent π terms,

$$Z_1 = f\left(\pi_1, \pi_2, \pi_3, \pi_4, \pi_5, \pi_6\right)$$

$$Z_2 = f\left(\pi_1, \pi_2, \pi_3, \pi_4, \pi_5, \pi_6\right) \qquad (6.1)$$

$$Z_3 = f\left(\pi_1, \pi_2, \pi_3, \pi_4, \pi_5, \pi_6\right)$$

f stands for "function of "
   Where
   $Z_1 = \left(\pi_{D1}\right)$, first dependent π term $= Td * R/Dr$
   $Z_2 = \left(\pi_{D2}\right)$, second dependent π term $= Pd * Dr/R$
   $Z_3 = \left(\pi_{D3}\right)$, third dependent π term $= He/Dr^3 * Pa$

To evolve the probable mathematical form of dimensional Equation 6.1 of the phenomenon, we can assume that the relationship among the dependent and independent π terms to be of exponential form. We can write the exact mathematical form of general equation as follows:

$$\pi_D = k*\left[\left(\pi_1\right)^a *\left(\pi_2\right)^b *\left(\pi_3\right)^c *\left(\pi_4\right)^d *\left(\pi_5\right)^e *\left(\pi_6\right)^f\right] \qquad (6.2)$$

This equation represents the following independent π terms involved in the face drilling phenomena.

$$\pi_D = K*\left[\left(N * A_2 * A_4 * A_6 * A_1\right)/\left(A_m * A_3 * A_5 * A_7\right)\right]^a, \left[\left(Lr * DC * Lc\right)/\left(Dr^3\right)\right]^b,$$

$$\left[\left\{\left(Wr * Wc * Wj\right)/\left(Dr^2 * Pa\right)^3\right\} * \left\{\left(So * Ss * Hr\right)/\left(Pa\right)^3\right\} * \left\{\left(Do * R^2\right)/Pa\right\}\right.$$

$$\left. * \left\{\left(Ds * R^2\right)/\left(Pa\right)\right\}\right]^c,$$

$$\left[\left\{\left(N * Dr * Ar\right)/\left(R\right)^2\right\} * \left\{\left(Qw/Dr^2 * R\right)\right\}\right]^d, \left[\left(\theta\right) * \left(\phi\right)\right]^e, \left[I/\left(Pa * R\right)\right]^f \qquad (6.3)$$

Equation 6.1 contains seven unknowns, the curve-fitting constant K, and the indices a, b, c, d, e, and f. To get the values of these unknowns, we need a minimum of seven sets of values of π_1, π_2, π_3, π_4, π_5, π_6, and π_D.

According to the plan in the design of field study, we have 30 sets of such values as given in Appendix Tables 1–9. If any arbitrary seven sets are chosen from this table, the values of the unknowns, K, a, b, c, d, e, and f are computed. One may not have a unique solution that represents the best curve fit for the remaining sets of values. To be very specific, one can select n number of combinations of r number of sets ^nC_r chosen out of n number of sets of values or $^{30}C_7$ in our case. Solving these many sets and finding their solutions is a herculean task. Hence, we employ the curve-fitting technique [22]. To implement this method to our field study, it is imperative to have the equation in the following form:

$$Y = a + bV + cW + dX\ldots \tag{6.4}$$

By taking log on both sides of Equation 6.1, the above mentioned form of equation can be obtained as shown below

$$\text{Log } \pi_D = \text{Log } k + a * \text{Log } (\pi_1) + b * \text{Log}(\pi_2) + c * \text{Log}(\pi_3)$$

$$+ d * \text{Log}(\pi_4) + e * \text{Log}(\pi_5) + f * \text{Log}(\pi_6) \tag{6.5}$$

Let us denote

$$\text{Log } \pi_D = Z,$$

$$\text{Log } k = K$$

$$\text{Log } (\pi_1) = A$$

$$\text{Log}(\pi_2) = B$$

$$\text{Log}(\pi_3) = C$$

$$\text{Log}(\pi_4) = D$$

$$\text{Log}(\pi_5) = E$$

$$\text{Log}(\pi_6) = F$$

Now, Equation 6.4 can be written as follows:

$$Z = K + a * A + b * B + c * C + d * D + e * E = f * F \tag{6.6}$$

Equation 6.5 is a regression equation of Z on A, B, C, D, E, and F in an N-dimensional co-ordinate system. This represents a regression hyper-plane. Substituting the data collected from the field, one would get such equation for every miner or 30 such equations in this case of 30 miners. The curve-fitting method, to include every set of readings, instead of any seven randomly chosen sets of readings, instructs us to add all these equations [27–29]. This results in the following equation:

$$\sum Z = nK + a * \sum A + b * \sum B + c * \sum C + d * \sum D + e * \sum E + f * \sum F$$

$$\tag{6.7}$$

Number of readings, n is 30.

As mentioned earlier, there are seven unknowns in this equation, K, a, b, c, d, e, and f. Hence, we will require seven different equations to solve. To achieve these equations, we multiply Equation 6.6 with A, B, C, D, E, and F to get a set of seven different equations as shown below:

$$\sum Z*A = K*\sum A + a*\sum A*A + b*\sum B*A + c*\sum C*A + d*\sum D*A$$
$$+ e*\sum E*A + f*\sum F*A$$

$$\sum Z*B = K*\sum B + a*\sum A*B + b*\sum B*B + c*\sum C*B + d*\sum D*B$$
$$+ e*\sum E*B + f*\sum F*B$$

$$\sum Z*C = K*\sum C + a*\sum A*C + b*\sum B*C + c*\sum C*C + d*\sum D*C$$
$$+ e*\sum E*C + f*\sum F*C$$

$$\sum Z*D = K*\sum D + a*\sum A*D + b*\sum B*D + c*\sum C*D + d*\sum D*D$$
$$+ e*\sum E*D + f*\sum F*D$$

$$\sum Z*E = K*\sum E + a*\sum A*E + b*\sum B*E + c*\sum C*E + d*\sum D*E$$
$$+ e*\sum E*E + f*\sum F*E$$

$$\sum Z*F = K*\sum F + a*\sum A*F + b*\sum B*F + c*\sum C*F + d*\sum D*F$$
$$+ e*\sum E*F + f*\sum F*F$$

Including Equation 6.6, we have a set of seven equations for the seven unknowns. To solve these equations, we shall use Matrix analysis. The Matrix format of an equation is of the form.

$$Z = W \times X$$

In the above mentioned equation, **X** is the unknown matrix. We solve this equation to determine the value of **X** by taking an inverse of matrix **W**. The final equation solved by using MATLAB is as follows.

$$X = inv(W) \times Z$$

Solving this equation for every dependent π term, the model is formed.

6.1.1 MODEL FORMULATION FOR TIME IN FACE DRILLING OPERATION BY IDENTIFYING THE CURVE-FITTING CONSTANT AND VARIOUS INDICES OF π TERMS

The multiple regression analysis helps to identify the indices of the different π terms in the model aimed at, by considering six independent π terms and one dependent π term. Let model aimed at be of the form,

$$(Z_1) = K_1 * \left[(\pi_1)^{a1} * (\pi_2)^{b1} * (\pi_3)^{c1} * (\pi_4)^{d1} * (\pi_5)^{e1} * (\pi_6)^{f1} \right] \tag{6.8}$$

To determine the values of a_1, b_1, c_1, d_1, e_1, and f_1 and to arrive at the regression hyper plane, the above mentioned equations are presented as follows:

$$\sum Z_1 = nK_1 + a_1 * \sum A + b_1 * \sum B + c_1 * \sum C + d_1 * \sum D + e_1 * \sum E + f_1 * \sum F$$

$$\sum Z_1 * A = K_1 * \sum A + a_1 * \sum A * A + b_1 * \sum B * A + c_1 * \sum C * A$$
$$+ d_1 * \sum D * A + e_1 * \sum E * A + f_1 * \sum F * A$$

$$\sum Z_1 * B = K_1 * \sum B + a_1 * \sum A * B + b_1 * \sum B * B + c_1 * \sum C * B$$
$$+ d_1 * \sum D * B + e_1 * \sum E * B + f_1 * \sum F * B$$

$$\sum Z_1 * C = K_1 * \sum C + a_1 * \sum A * C + b_1 * \sum B * C + c_1 * \sum C * C$$
$$+ d_1 * \sum D * C + e_1 * \sum E * C + f_1 * \sum F * C$$

$$\sum Z_1 * D = K_1 * \sum D + a_1 * \sum A * D + b_1 * \sum B * D + c_1 * \sum C * D$$
$$+ d_1 * \sum D * D + e_1 * \sum E * D + f_1 * \sum F * D$$

$$\sum Z_1 * E = K_1 * \sum E + a_1 * \sum A * E + b_1 * \sum B * E + c_1 * \sum C * E$$
$$+ d_1 * \sum D * E + e_1 * \sum E * E + f_1 * \sum F * E$$

$$\sum Z_1 * F = K_1 * \sum F + a_1 * \sum A * F + b_1 * \sum B * F + c_1 * \sum C * F$$
$$+ d_1 * \sum D * F + e_1 * \sum E * F + f_1 * \sum F * F$$

After substituting these values in the equations, one would get a set of seven equations, which are to be solved simultaneously to get the values of K_1, a_1, b_1, c_1, d_1, e_1, and f_1. The above mentioned equations can be verified in the matrix form and further values of K_1, a_1, b_1, c_1, d_1, e_1, and f_1 can be obtained by using matrix analysis.

$$Z_1 = W_1 \times X_1$$

Here,

$W_1 = 7 \times 7$ matrix of the multipliers of K_1, a_1, b_1, c_1, d_1, e_1, and f_1
$P_1 = 7 \times 1$ matrix of the terms on L H S and
$X_1 = 7 \times 1$ matrix of solutions of values of K_1, a_1, b_1, c_1, d_1, e_1, and f_1

Then, the matrix obtained is given by,
Matrix

$$
\begin{bmatrix} Z_1 \\ A \\ B \\ C \\ D \\ E \\ F \end{bmatrix}
=
\begin{bmatrix}
n & A & B & C & D & E & F \\
A & A^2 & BA & CA & DA & EA & FA \\
B & AB & B^2 & CB & DB & EB & FB \\
C & AC & BC & C^2 & DC & EC & FC \\
D & AD & BD & CD & D^2 & ED & FD \\
E & AE & BE & CE & DE & E^2 & FE \\
F & AF & BF & CF & DF & EF & F^2
\end{bmatrix}
\begin{bmatrix} K_1 \\ a_1 \\ b_1 \\ c_1 \\ d_1 \\ e_1 \\ f_1 \end{bmatrix}
$$

In the above mentioned equations, n is the number of sets of readings, A, B, C, D, E, and F represent the independent π terms π_1, π_2, π_3, π_4, π_5 and π_6 while Z_1 matrix represents dependent π term.

Substituting the values of A, A^2, BA, CA........up to F^2 in the abovementioned matrix (Refer Annexure 12.1 to 12.9), value of constant K_1 and indices a_1, b_1, c_1, d_1, e_1, and f_1 are evaluated by taking the inverse of matrix W_1 and multiplying with Z_1 matrix as shown

$$
\begin{bmatrix}
43.547 \\ -53.006 \\ -161.091 \\ 34.887 \\ 1.239 \\ 60.895 \\ 9.752
\end{bmatrix}
=
\begin{bmatrix}
30 & -36.494 & -110.519 & 23.554 & 0.814 & 41.910 & 6.726 \\
-36.494 & 45.273 & 134.430 & -28.657 & -0.961 & -50.997 & -8.214 \\
-110.51 & 134.43 & 408.08 & -87.632 & -3.286 & -154.44 & -24.759 \\
23.554 & -28.657 & -87.632 & 19.342 & 0.884 & 32.961 & 5.264 \\
0.814 & -0.961 & -3.286 & 0.884 & 0.253 & 1.142 & 0.177 \\
41.910 & -50.997 & -154.441 & 32.961 & 1.142 & 58.55 & 9.395 \\
6.726 & -8.214 & -24.759 & 5.264 & 0.177 & 9.395 & 1.520
\end{bmatrix}
$$

$$
\times
\begin{bmatrix}
K_1 \\ a_1 \\ b_1 \\ c_1 \\ d_1 \\ e_1 \\ f_1
\end{bmatrix}
$$

$K_1 = 0.6025$, $a_1 = 0.0193$, $b_1 = 0.0312$, $c_1 = 1.0528$, $d_1 = -0.8358$, $e_1 = 0.6797$, and $f_1 = 0.2519$

The exact form of model obtained is as follows:

$$(Z_1) = 0.6025 * (\pi_1)^{0.0193} * (\pi_2)^{0.0312} * (\pi_3)^{1.0528} * (\pi_4)^{-0.8358} * (\pi_5)^{0.6797} * (\pi_6)^{0.2515}$$

(6.9)

Time for Face Drilling (Z_1) =

$$\left(Td * R/Dr\right) = 0.6025 * \left[\left\{(N * A_2 * A_4 * A_6 * A_1)/(A_m * A_3 * A_5 * A_7)\right\}\right]^{0.0193},$$

$$\left[(Lr * DC * Lc)/(Dr^3)\right]^{0.0312},$$

$$\left[\left\{(Wr * Wc * Wj)/(Dr^2 * Pa)^3\right\} * \left\{(So * Ss * Hr)/(Pa)^3\right\} * \left\{(Do * R^2)/Pa\right\} *\right.$$

$$\left.\left\{(Ds * R^2)/(Pa)\right\}\right]^{1.0528},$$

$$\left[\left\{(Dr * N * Ar)/(R)^2\right\} * \left\{(Qw/Dr^2 * R)\right\}\right]^{-0.8358}, \left[(\theta) * (\phi)\right]^{0.6797}, \left[I/(R * Pa)\right]^{0.2519}$$

(6.10)

6.1.2 Model Formulation for Productivity in Face Drilling Operation by Identifying the Curve-Fitting Constant and Various Indices of π Terms

The multiple regression analysis helps to identify the indices of different π terms in the model aimed at, by considering six independent π terms and one dependent π term. Let model aimed at be of the form,

$$(Z_2) = K_2 * \left[(\pi_1)^{a2} * (\pi_2)^{b2} * (\pi_3)^{c2} * (\pi_4)^{d2} * (\pi_5)^{e2} * (\pi_6)^{f2}\right] \qquad (6.11)$$

To determine the values of a_2, b_2, c_2, d_2, e_2, and f_2 and to arrive at the regression hyper plane, the above equations are presented as follows:

$$\sum Z_2 = nK_2 + a_2 * \sum A + b_2 * \sum B + c_2 * \sum C + d_2 * \sum D + e_2 * \sum E + f_2 * \sum F$$

$$\sum Z_2 * A = K_2 * \sum A + a_2 * \sum A * A + b_2 * \sum B * A + c_2 * \sum C * A$$

$$+ d_2 * \sum D * A + e_2 * \sum E * A + f_2 * \sum F * A$$

$$\sum Z_2 * B = K_2 * \sum B + a_2 * \sum A * B + b_2 * \sum B * B + c_2 * \sum C * B$$

$$+ d_2 * \sum D * B + e_2 * \sum E * B + f_2 * \sum F * B$$

$$\sum Z_2 * C = K_2 * \sum C + a_2 * \sum A * C + b_2 * \sum B * C + c_2 * \sum C * C$$

$$+ d_2 * \sum D * C + e_2 * \sum E * C + f_2 * \sum F * C$$

$$\sum Z_2 * D = K_2 * \sum D + a_2 * \sum A * D + b_2 * \sum B * D + c_2 * \sum C * D$$

$$+ d_2 * \sum D * D + e_2 * \sum E * D + f_2 * \sum F * D$$

$$\sum Z_2 * E = K_2 * \sum E + a_2 * \sum A * E + b_2 * \sum B * E + c_2 * \sum C * E$$

$$+ d_2 * \sum D * E + e_2 * \sum E * E + f_2 * \sum F * E$$

$$\sum Z_2 * F = K_2 * \sum F + a_2 * \sum A * F + b_2 * \sum B * F + c_2 * \sum C * F$$

$$+ d_2 * \sum D * F + e_2 * \sum E * F + f_2 * \sum F * F$$

In the above set of equations, the values of the multipliers K_2, a_2, b_2, c_2, d_2, e_2, and f_2 are substituted to compute the values of the unknowns (viz. K_2, a_2, b_2, c_2, d_2, e_2, and f_2). The values of the terms on L.H.S. and the multipliers of K_2, a_2, b_2, c_2, d_2, e_2, and f_2 in the set of equations are calculated. After substituting these values in the equations, one will get a set of seven equations, which are to be solved simultaneously to get the values of K_2, a_2, b_2, c_2, d_2, e_2, and f_2. The above equations can be verified in the matrix form and further values of K_2, a_2, b_2, c_2, d_2, e_2, and f_2 can be obtained by using matrix analysis.

$$Z_2 = W_2 \times X_2$$

Here,
   $W_2 = 7 \times 7$ matrix of the multipliers of K_2, a_2, b_2, c_2, d_2, e_2, and f_2
   $P_2 = 7 \times 1$ matrix of the terms on LHS and
   $X_2 = 7 \times 1$ matrix of solutions of values of K_2, a_2, b_2, c_2, d_2, e_2, and f_2

Then, the matrix obtained is given by

Matrix

$$
\begin{bmatrix} Z_2 \\ A \\ B \\ C \\ D \\ E \\ F \end{bmatrix}
=
\begin{bmatrix}
n & A & B & C & D & E & F \\
A & A^2 & BA & CA & DA & EA & FA \\
B & AB & B^2 & CB & DB & EB & FB \\
C & AC & BC & C^2 & DC & EC & FC \\
D & AD & BD & CD & D^2 & ED & FD \\
E & AE & BE & CE & DE & E^2 & FE \\
F & AF & BF & CF & DF & EF & F^2
\end{bmatrix}
\begin{bmatrix} K_2 \\ a_2 \\ b_2 \\ c_2 \\ d_2 \\ e_2 \\ f_2 \end{bmatrix}
$$

In the above equations, n is the number of sets of readings, A, B, C, D, E, and F represent the independent π terms π_1, π_2, π_3, π_4, π_5 and π_6 while, Z_2 matrix represents, dependent π term.

Substituting the values of A, A^2, BA, CA........up to F^2 in the above matrix (Refer Annexure 13.1 to 13.9), value of constant K_2 and indices a_2, b_2, c_2, d_2, e_2, and f_2 are evaluated by taking the inverse of matrix W_2 and multiplying with Z_2 matrix as shown

$$
\begin{bmatrix} 8.807 \\ -10.726 \\ -32.467 \\ 6.921 \\ 0.281 \\ 12.300 \\ 1.979 \end{bmatrix}
=
\begin{bmatrix}
30 & -36.494 & -110.519 & 23.554 & 0.814 & 41.910 & 6.726 \\
-36.494 & 45.273 & 134.430 & -28.657 & -0.961 & -50.997 & -8.214 \\
-110.51 & 134.43 & 408.08 & -87.632 & -3.286 & -154.44 & -24.759 \\
23.554 & -28.657 & -87.632 & 19.342 & 0.884 & 32.961 & 5.264 \\
0.814 & -0.961 & -3.286 & 0.884 & 0.253 & 1.142 & 0.177 \\
41.910 & -50.997 & -154.441 & 32.961 & 1.142 & 58.55 & 9.395 \\
6.726 & -8.214 & -24.759 & 5.264 & 0.177 & 9.395 & 1.520
\end{bmatrix}
$$

$$
\times
\begin{bmatrix} K_2 \\ a_2 \\ b_2 \\ c_2 \\ d_2 \\ e_2 \\ f_2 \end{bmatrix}
$$

$K_2 = 0.1706$, $a_2 = -0.0056$, $b_2 = -0.2037$, $c_2 = -0.2735$, $d_2 = 0.2250$, $e_2 = 0.2933$, and $f_2 = 0.4595$

The exact form of model obtained is as follows:

$$
(Z_2) = 0.1706 * (\pi_1)^{-0.0056} * (\pi_2)^{-0.2037} * (\pi_3)^{-0.2735} * (\pi_4)^{0.2250} * (\pi_5)^{0.2933} * (\pi_6)^{0.4595}
$$

$$(6.12)$$

Productivity for Face Drilling (Z_2) =

$$\left(\text{Pd}/\text{Dr} * \text{R}\right) = 0.1706 * \left[\left\{\left(\text{N} * \text{A}_2 * \text{A}_4 * \text{A}_6 * \text{A}_1\right)/\left(\text{A}_m * \text{A}_3 * \text{A}_5 * \text{A}_7\right)\right\}\right]^{-0.0056},$$

$$\left[\left(\text{Lr} * \text{DC} * \text{Lc}\right)/\left(\text{Dr}^3\right)\right]^{-0.2037},$$

$$\left[\left\{\left(\text{Wr} * \text{Wc} * \text{Wj}\right)/\left(\text{Dr}^2 * \text{Pa}\right)^3\right\} * \left\{\left(\text{So} * \text{Ss} * \text{Hr}\right)/\left(\text{Pa}\right)^3\right\} * \left\{\left(\text{Do} * \text{R}^2\right)/\text{Pa}\right\} *\right.$$

$$\left.\left\{\left(\text{Ds} * \text{R}^2\right)/\left(\text{Pa}\right)\right\}\right]^{-0.2735},$$

$$\left[\left\{\left(\text{Dr} * \text{N} * \text{Ar}\right)/\left(\text{R}\right)^2\right\} * \left\{\left(\text{Qw}/\text{Dr}^2 * \text{R}\right)\right\}\right]^{0.2250}, \left[\left(\theta\right) * \left(\phi\right)\right]^{0.2933}, \left[\text{I}/\left(\text{R} * \text{Pa}\right)\right]^{0.4595}$$

$$(6.13)$$

6.1.3 MODEL FORMULATION FOR HUMAN ENERGY CONSUMED IN FACE DRILLING OPERATION BY IDENTIFYING THE CURVE-FITTING CONSTANT AND VARIOUS INDICES OF π TERMS

The multiple regression analysis helps to identify the indices of the different π terms in the model aimed at, by considering six independent π terms and one dependent π term. Let model aimed at be of the form,

$$\left(Z_3\right) = K_3 * \left[\left(\pi_1\right)^{a3} * \left(\pi_2\right)^{b3} * \left(\pi_3\right)^{c3} * \left(\pi_4\right)^{d3} * \left(\pi_5\right)^{e3} * \left(\pi_6\right)^{f3}\right] \qquad (6.14)$$

To determine the values of a_3, b_3, c_3, d_3, e_3, and f_3 and to arrive at the regression hyper plane, the above equations are presented as follows:

$$\sum Z_3 = nK_3 + a_3 * \sum A + b_3 * \sum B + c_3 * \sum C + d_3 * \sum D + e_3 * \sum E$$

$$\sum Z_3 * A = K_3 * \sum A + a_3 * \sum A * A + b_3 * \sum B * A + c_3 * \sum C * A$$
$$+ d_3 * \sum D * A + e_3 * \sum E * A$$

$$\sum Z_3 * B = K_3 * \sum B + a_3 * \sum A * B + b_3 * \sum B * B + c_3 * \sum C * B$$
$$+ d_3 * \sum D * B + e_3 * \sum E * B$$

$$\sum Z_3 * C = K_3 * \sum C + a_3 * \sum A * C + b_3 * \sum B * C + c_3 * \sum C * C$$
$$+ d_3 * \sum D * C + e_3 * \sum E * C$$

$$\sum Z_3 * D = K_3 * \sum D + a_3 * \sum A * D + b_3 * \sum B * D + c_3 * \sum C * D$$

$$+ d_3 * \sum D * D + e_3 * \sum E * D$$

$$\sum Z_3 * E = K_3 * \sum E + a_3 * \sum A * E + b_3 * \sum B * E + c_3 * \sum C * E$$

$$+ d_3 * \sum D * E + e_3 * \sum E * E$$

In the above set of equations, the values of the multipliers K_3, a_3, b_3, c_3, d_3, e_3, and f_3 are substituted to compute the values of the unknowns (viz., K_3, a_3, b_3, c_3, d_3, e_3, and f_3). The values of the terms on LHS and the multipliers of K_3, a_3, b_3, c_3, d_3, e_3, and f_3 in the set of equations are calculated. After substituting these values in the equations, one would get a set of seven equations, which are to be solved simultaneously to get the values of K_3, a_3, b_3, c_3, d_3, e_3, and f_3. The above equations can be verified in the matrix form and further values of K_3, a_3, b_3, c_3, d_3, e_3, and f_3 can be obtained by using matrix analysis.

$$\mathbf{Z_3 = W_3 \times X_3}$$

Here,
$\mathbf{W_3} = 7 \times 7$ matrix of the multipliers of K_3, a_3, b_3, c_3, d_3, e_3, and f_3
$\mathbf{P_3} = 7 \times 1$ matrix of the terms on L H S and
$\mathbf{X_3} = 7 \times 1$ matrix of solutions of values K_3, a_3, b_3, c_3, d_3, e_3, and f_3
Then, the matrix obtained is given by,
Matrix

$$
\begin{bmatrix}
Z_3 \\
A \\
B \\
C \\
D \\
E \\
F
\end{bmatrix}
=
\begin{bmatrix}
n & A & B & C & D & E & F \\
A & A^2 & BA & CA & DA & EA & FA \\
B & AB & B^2 & CB & DB & EB & FB \\
C & AC & BC & C^2 & DC & EC & FC \\
D & AD & BD & CD & D^2 & ED & FD \\
E & AE & BE & CE & DE & E^2 & FE \\
F & AF & BF & CF & DF & EF & F^2
\end{bmatrix}
\begin{bmatrix}
K_3 \\
a_3 \\
b_3 \\
c_3 \\
d_3 \\
e_3 \\
f_3
\end{bmatrix}
$$

In the above equations, n is the number of sets of readings, A, B, C, D, E, and F represent the independent π terms π_1, π_2, π_3, π_4, π_5 and π_6 while, Z_3 matrix represents, dependent π term.

Substituting the values of A, A^2, BA, CA……..up to F^2 in the above matrix (Refer Annexure 14.1 to 14.9), value of constant K_3 and indices a_3, b_3, c_3, d_3, e_3, and f_3 are evaluated by taking the inverse of matrix W_3 and multiplying with Z_3 matrix as shown

$$
\begin{bmatrix} 25.903 \\ -31.554 \\ -96.245 \\ 21.114 \\ 0.882 \\ 36.242 \\ 5.813 \end{bmatrix} = \begin{bmatrix} 30 & -36.494 & -110.519 & 23.554 & 0.814 & 41.910 & 6.726 \\ -36.494 & 45.273 & 134.430 & -28.657 & -0.961 & -50.997 & -8.214 \\ -110.51 & 134.43 & 408.08 & -87.632 & -3.286 & -154.44 & -24.759 \\ 23.554 & -28.657 & -87.632 & 19.342 & 0.884 & 32.961 & 5.264 \\ 0.814 & -0.961 & -3.286 & 0.884 & 0.253 & 1.142 & 0.177 \\ 41.910 & -50.997 & -154.441 & 32.961 & 1.142 & 58.55 & 9.395 \\ 6.726 & -8.214 & -24.759 & 5.264 & 0.177 & 9.395 & 1.520 \end{bmatrix}
$$

$$
\times \begin{bmatrix} K_3 \\ a_3 \\ b_3 \\ c_3 \\ d_3 \\ e_3 \\ f_3 \end{bmatrix}
$$

$K_3 = 0.000319$, $a_3 = 0.0495$, $b_3 = -0.5893$, $c_3 = 0.4369$, $d_3 = -0.3944$, $e_3 = 1.0414$, and $f_3 = 2.0566$

The exact form of model obtained is as follows:

$$
(Z_3) = 0.000319 * (\pi_1)^{0.0495} * (\pi_2)^{-0.5893} * (\pi_3)^{0.4369} * (\pi_4)^{-0.3944} * (\pi_5)^{1.0414} * (\pi_6)^{2.0566}
$$

(6.15)

Human Energy consumed for Face Drilling (Z_3) =

$$
(He/Dr^3 * Pa) = 0.000319 * \left[\left\{ (N * A_2 * A_4 * A_6 * A_1)/(A_m * A_3 * A_5 * A_7) \right\} \right]^{0.0495},
$$

$$
\left[(Lr * DC * Lc)/(Dr^3) \right]^{-0.5893},
$$

$$
\left[\left\{ (Wr * Wc * Wj)/(Dr^2 * Pa)^3 \right\} * \left\{ (So * Ss * Hr)/(Pa)^3 \right\} \left\{ (Do * R^2)/Pa \right\} * \right.
$$

$$
\left. \left\{ (Ds * R^2)/(Pa) \right\} \right]^{0.4369},
$$

$$
\left[\left\{ (Dr * N * Ar)/(R)^2 \right\} * \left\{ (Qw/Dr^2 * R) \right\} \right]^{-0.3944}, \left[(\theta) * (\phi) \right]^{1.0414}, \left[I/(R * Pa) \right]^{2.0566}
$$

(6.16)

6.2 MODELS DEVELOPED FOR THE DEPENDENT VARIABLES-FACE DRILLING

The exact forms of models obtained for the dependent variables time, productivity and human energy consumed in face drilling operation are as follows:

$$(Z_1) = 0.6025 * (\pi_1)^{0.0193} * (\pi_2)^{0.0312} * (\pi_3)^{1.0528} * (\pi_4)^{-0.8358} * (\pi_5)^{0.6797} * (\pi_6)^{0.2519}$$

$$(6.9)$$

$$(Z_2) = 0.1706 * (\pi_1)^{-0.0056} * (\pi_2)^{-0.2037} * (\pi_3)^{-0.2735} * (\pi_4)^{0.2250} * (\pi_5)^{0.2933} * (\pi_6)^{0.4595}$$

$$(6.12)$$

$$(Z_3) = 0.000319 * (\pi_1)^{0.0495} * (\pi_2)^{-0.5893} * (\pi_3)^{0.4369} * (\pi_4)^{-0.3944} * (\pi_5)^{1.0414} * (\pi_6)^{2.0566}$$

$$(6.15)$$

Interpretation of models is detailed in Chapter 9.

7 Artificial Neural Network Simulation

Pramod Belkhode
Laxminarayan Institute of Technology

Sarika Modak
Priyadarshini College of Engineering

Vinod Ganvir and Anand Shende
Laxminarayan Institute of Technology

CONTENTS

7.1   INTRODUCTION

The field data-based modeling has been achieved based on field data for the three dependent π terms for face drilling. In such complex phenomenon involving the validation of field data-based models is not in close proximity, it becomes necessary to formulate artificial neural network (ANN) simulation of the observed data. Simulation consists of three layers [30–31]. The first layer is known as input layer. The number of independent variables is equal to the number of neurons in input layer. The hidden layer is the second layer. The output layer is the third layer. It contains one neuron as one of dependent variables.

7.2   PROCEDURE FOR FORMATION OF ANN SIMULATION

ANN simulation is run on the MATLAB software. The steps are as follows.

1. The observed data from the field study is the input data and output data are imported to the program, respectively.
2. The input and output data are read by prestd function and appropriately sized. Function prestd preprocesses the data so that the mean is 0 and standard deviation is 1.

3. In the preprocessing step, the input and output data are normalized using mean and standard deviation.
4. The input and output data are then categorized into three categories, namely, testing, validation, and training. From the 30 observations for face drilling and 30 observations for manual loading, initial 75% of the observations for both is selected for training, last 75% data for validation, and middle overlapping 50% data for testing.
5. The data are then stored in structures for training, testing, and validation.
6. Looking at the pattern of the data, feed forward back propagation-type neural network is chosen.
7. This network is then trained using the training data.
8. The regression analysis and the representation are performed through the standard functions. The values of regression coefficient and the equation of regression lines are represented on one different graph plotted for the one dependent π term.

7.3   ANN PROGRAM

MATLAB software is selected for developing ANN. The program executed for face drilling operation is given as follows (Figures 7.1 to 7.8):

7.3.1   ANN PROGRAM FOR FACE DRILLING – TIME

```
clear all;
close all;
inputs3=[
```

0.039187621	0.000517158	2.9403891	0.823957	24.3	1.652891
0.036227419	0.000369398	3.4835941	0.84540773	24.3	1.74394
0.026762664	0.000323224	3.8087295	0.98897029	24.3	1.843125
0.152285379	0.00028731	4.5611273	1.04933096	24.57	1.551045
0.088862089	0.000258579	4.6687068	1.40987695	24.57	1.673048
0.082202885	0.000315237	4.9725657	0.84892539	25.48	1.652891
0.098373217	0.000225169	6.3456155	0.94149989	25.48	1.711843
0.069134398	0.000197023	7.3191638	1.38870586	25.76	1.785165
0.057247864	0.000175132	7.1950756	1.10343967	25.76	1.578256
0.054169457	0.000157618	8.9031511	1.52857684	25.76	1.62355
0.030461482	0.000236428	6.3066687	0.82062788	26.97	1.652891
0.059075717	0.000168877	8.2474219	0.99040943	26.97	1.701405
0.043114358	0.000147767	9.7587897	1.06619738	26.97	1.762989

0.087729067	0.000131349	8.9138659	1.12302691	25.81	1.58754
0.062446217	0.000118214	10.766692	1.37571451	25.81	1.642993
0.096785416	0.000430965	2.9403891	0.823957	24.03	1.652891
0.071433543	0.000323224	3.4835941	0.84540773	23.76	1.74394
0.071339852	0.00028731	3.8087295	0.98897029	23.76	1.843125
0.049634396	0.000258579	4.5611273	1.04933096	23.76	1.551045
0.056999177	0.000215482	4.6687068	1.40987695	23.49	1.673048
0.04884502	0.000262697	4.9725657	0.84892539	24.36	1.652891
0.03871031	0.000197023	6.3456155	0.94149989	24.36	1.711843
0.052611517	0.000175132	7.3191638	1.38870586	24.08	1.785165
0.068567615	0.000157618	7.1950756	1.10343967	24.08	1.578256
0.08350799	0.000131349	8.9031511	1.52857684	24.08	1.62355
0.127091286	0.000197023	6.3066687	0.82062788	24.65	1.652891
0.074060057	0.000147767	8.2474219	0.99040943	24.65	1.701405
0.051763038	0.000131349	9.7587897	1.06619738	24.65	1.762989
0.049590511	0.000118214	8.9138659	1.12302691	26.1	1.58754
0.044045203	9.85115E−05	10.766692	1.37571451	26.39	1.642993

```
]
a1=inputs3
a2=a1
input_data=a2;
output3=[
]


          17.57647059
          17.64705882
          17.66470588
          17.65588235
          17.60588235
          30.18181818
          30.37727273
          30.35454545
          30.31363636
          30.21515152
          42.25454545
          42.49090909
          42.44545455
          42.5
          42.25606061
          17.57647059
          17.64705882
          17.66470588
```

```
           17.65588235
           17.60588235
           30.18181818
           30.37727273
           30.35454545
           30.31363636
           30.21515152
           42.25454545
           42.49090909
           42.44545455
           42.5
           42.25606061
```

```
y1=output3
y2=y1
size(a2);
size(y2);
p=a2';
sizep=size(p);
t=y2';
sizet=size(t);
[S Q]=size(t)
[pn, meanp, stdp, tn, meant, stdt] = prestd(p, t);
net = newff(minmax(pn), [30 1], {'logsig' 'purelin'},
'trainlm');
net.performFcn='mse';
net.trainParam.goal=.99;
net.trainParam.show=200;
net.trainParam.epochs=50;
net.trainParam.mc=0.05;
net = train(net, pn, tn);
an = sim(net, pn);
[a] = poststd(an, meant, stdt);
error=t-a;
x1=1:30;
plot(x1,t, 'rs-', x1,a, 'b-')
legend('Experimental', 'Neural');
title('Output (Red) and Neural Network Prediction (Blue)
 Plot');
xlabel('Experiment No.');
ylabel('Output');
grid on;
figure
error_percentage=100*error./t
plot(x1,error_percentage)
legend('percentage error');
axis([0 30-100 100]);
title('Percentage Error Plot in Neural Network Prediction');
```

```
xlabel('Experiment No.');
ylabel('Error in %');
grid on;
for ii=1:30
xx1=input_data(ii, 1);
yy2=input_data(ii, 2);
zz3=input_data(ii, 3);
xx4=input_data(ii, 4);
yy5=input_data(ii, 5);
zz6=input_data(ii, 6);
pause
yyy(1,ii)=0.6025*power(xx1,0.0193)*power(yy2,0.0312)*power
  (zz3,1.0528) *power(xx4,-0.8358)*power(yy5,0.6797)*power
  (zz6,0.2519);
yy_practical(ii)=(y2(ii, 1));
yy_eqn(ii)=(yyy(1,ii))
yy_neur(ii)=(a(1,ii))
yy_practical_abs(ii)=(y2(ii, 1));
yy_eqn_abs(ii)=(yyy(1,ii));
yy_neur_abs(ii)=(a(1,ii));
pause
end
figure;
plot(x1,yy_practical_abs, 'r-', x1,yy_eqn_abs, 'b-', x1,yy_
  neur_abs, 'k-');
legend('Practical', 'Equation', 'Neural');
title('Comparison between practical data, equation-based data
  and neural-based data');
xlabel('Experimental');
grid on;
figure;
plot(x1,yy_practical_abs, 'r-', x1,yy_eqn_abs, 'b-');
legend('Practical',' Equation');
title('Comparison between practical data, equation-based data
  and neural-based data');
xlabel('Experimental');
grid on;
figure;
plot(x1,yy_practical_abs, 'r-', x1,yy_neur_abs, 'k-');
legend('Practical', 'Neural');
title('Comparison between practical data, equation-based data
  and neural-based data');
xlabel('Experimental');
grid on;
error1=yy_practical_abs-yy_eqn_abs
figure
error_percentage1=100*error1./yy_practical_abs;
plot(x1,error_percentage, 'k-', x1,error_percentage1,'b-');
legend('Neural', 'Equation');
axis([0 30-100 100]);
```

```
title('Percentage Error Plot in Equation (blue), Neural
 Network (black) Prediction');
xlabel('Experiment No.');
ylabel('Error in %');
grid on;
meanexp=mean(output3)
meanann=mean(a)
meanmath=mean(yy_eqn_abs)
mean_absolute_error_performance_function = mae(error)
mean_squared_error_performance_function = mse(error)
net = newff(minmax(pn), [30 1], {'logsig' 'purelin'},
 'trainlm', 'learngdm', 'msereg');
an = sim(net, pn);
[a] = poststd(an, meant, stdt);
error=t(1,[1:30])-a(1,[1:30]);
net.performParam.ratio = 20/(20+1);
perf = msereg(error, net)
rand('seed', 1.818490882E9)
[ps] = minmax(p);
[ts] = minmax(t);
numInputs = size(p, 1);
numHiddenNeurons = 30;
numOutputs = size(t, 1);
net = newff(minmax(p), [numHiddenNeurons, numOutputs]);
[pn, meanp, stdp, tn, meant, stdt] = prestd(p, t);
[ptrans, transmit]=prepca(pn, 0.001);
[R Q]=size(ptrans);
testSamples= 15:1:Q;
validateSamples=20:1:Q;
trainSamples= 1:1:Q;
validation.P=ptrans(:, validateSamples);
validation.T=tn(:, validateSamples);
testing.P= ptrans(:, testSamples);
testing.T= tn(:, testSamples)
ptr= ptrans(:, trainSamples);
ttr= tn(:, trainSamples);
net = newff(minmax(ptr), [30 1], {'logsig' 'purelin'},
'trainlm');
[net, tr] = train(net, ptr, ttr, [], [], validation, testing);
plot(tr.epoch, tr.perf, 'r', tr.epoch, tr.vperf, 'g',
 tr.epoch, tr.tperf, 'h');
legend('Training', 'validation', 'Testing', -1);
ylabel('Error');
an=sim(net, ptrans);
a=poststd(an, meant, stdt);
pause;
figure
[m, b, r] = postreg(a, t);
```

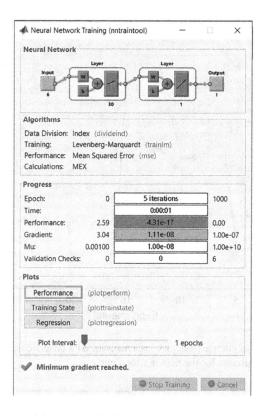

FIGURE 7.1 Training of the network for Time.

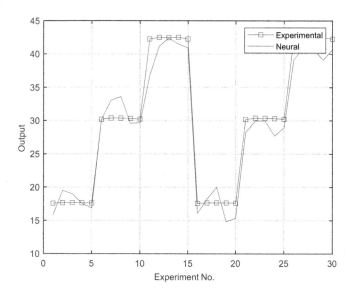

FIGURE 7.2 Graph of comparison with experimental data base and neural prediction for the network for time.

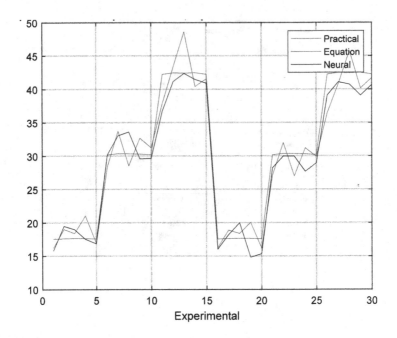

FIGURE 7.3 Graph of comparison with experimental data base, neural network prediction and equation base prediction for the network for Time.

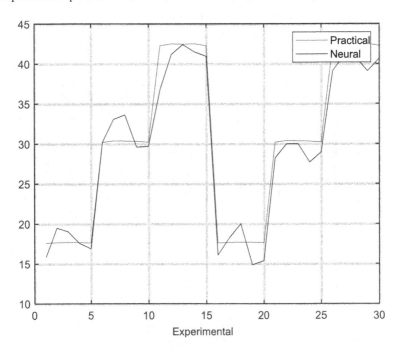

FIGURE 7.4 Graph of comparison with experimental data base and neural network prediction for the network for Time.

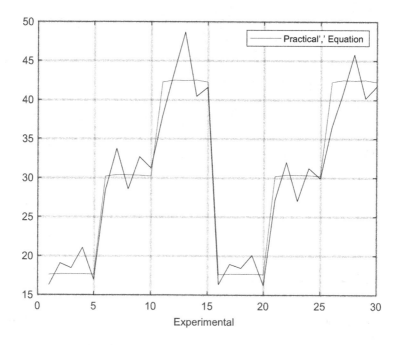

FIGURE 7.5   Graph of comparison of with mathematical equation base and experimental base for the network for Time.

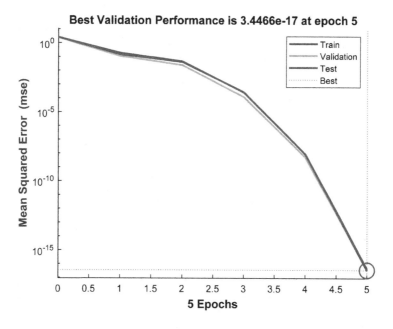

FIGURE 7.6   Graph of training of experimental data for the network for Time.

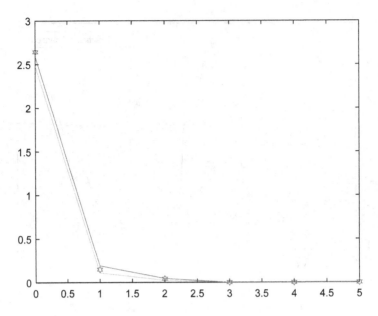

FIGURE 7.7 Graph of training, validation, testing of experimental data point for Time.

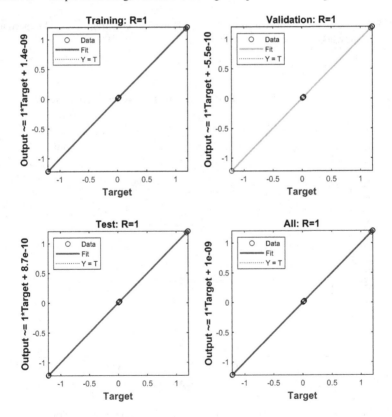

FIGURE 7.8 Graph regression analysis comparisons of actual and computed data by ANN for Time.

These graphs are a representation of the face drilling operation– Time using ANN Simulation

7.3.2 ANN PROGRAM FOR FACE DRILLING – PRODUCTIVITY

MATLAB program for the face drilling – productivity is same as the program for face drilling – Time except for the output for the face drilling – Productivity and the model equation which is different as shown below.

Output for the face drilling– Productivity

```
output3=[
```

        1.8489036
        2.0068500
        2.1464156
        1.9675057
        2.1594146
        1.7945241
        1.9119755
        2.0177736
        1.9431404
        2.0338935
        1.7945241
        1.9003171
        1.9927081
        1.9545706
        2.0582515
        1.8489036
        2.0068500
        2.1464156
        1.9675057
        2.1594146
        1.7945241
        1.9119755
        2.0177736
        1.9431404
        2.0338935
        1.7945241
        1.9003171
        1.9927081
        1.9545706
        2.0582515

```
]
```
Equation for the face drilling –Productivity

$$yyy(1,ii) = 0.1706 * power(xx1, -0.0056) * power(yy2, -0.2037)$$

$$* power(zz3, -0.2735) * power(xx4, 0.2250)$$

$$* power(yy5, 0.2933) * power(zz6, 0.4595);$$

The above output and equation of face drilling – productivity is placed instead of face drilling – time output and equation in MATLAB program of face drilling– time. Various output graphs obtained for face drilling–productivity are shown as below by the ANN program (Figures 7.9 to 7.16).

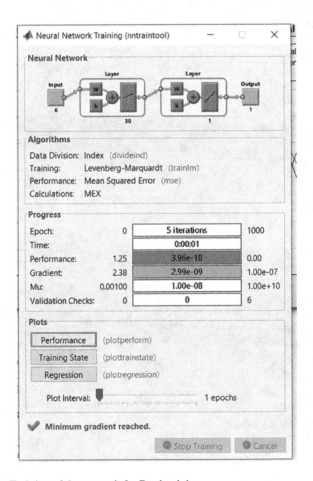

FIGURE 7.9   Training of the network for Productivity.

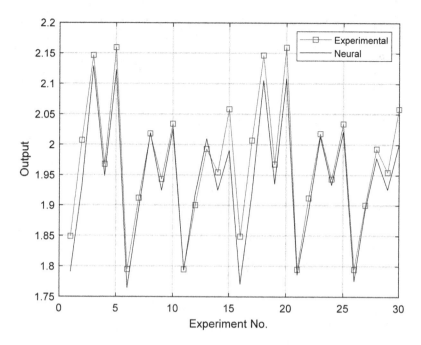

FIGURE 7.10 Graph of comparison with experimental data base and neural prediction for the network for Productivity.

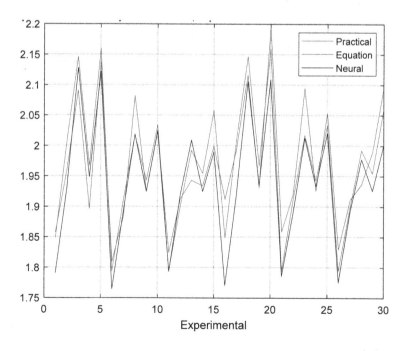

FIGURE 7.11 Graph of comparison with experimental data base, neural network prediction and equation base prediction for the network for Productivity.

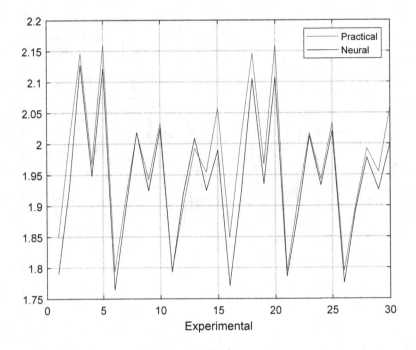

FIGURE 7.12 Graph of comparison with experimental data base and neural network base for the network for Productivity.

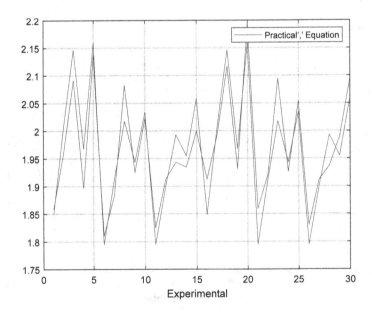

FIGURE 7.13 Graph of comparison with mathematical equation base and experimental data base for the network for Productivity.

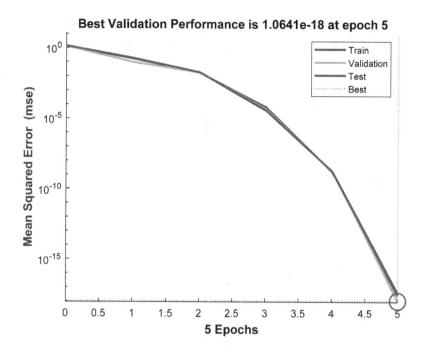

FIGURE 7.14   Graph of training of experimental data for the network for Productivity.

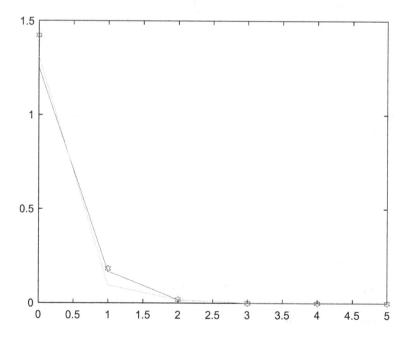

FIGURE 7.15   Graph of training, validation, testing of experimental data point for Productivity.

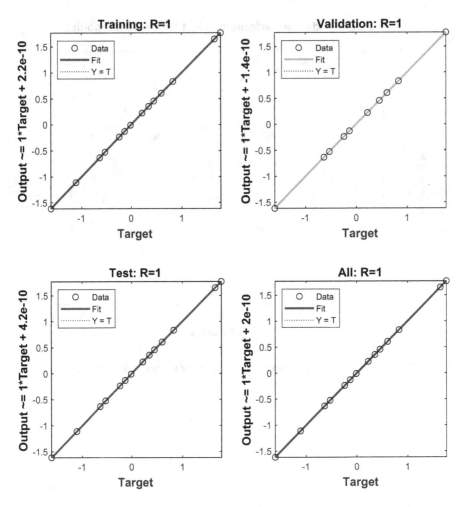

FIGURE 7.16   Graph regression analysis comparison of actual and computed data by ANN for Productivity.

These graphs are a representation of the face drilling operation– productivity using ANN simulation.

7.3.3   ANN Program for Face Drilling Operation – Human Energy

MATLAB program for the face drilling – human energy is the same as the program for face drilling– time except for the output for the face drilling– human energy.

Output for the face drilling– human energy

```
output3=[
```

3.256415279
5.985657551
5.091961407
3.066925439
5.539869779
8.806299006
5.800968168
8.822842891
8.174471432
9.943080914
6.851307189
8.33913453
12.12506649
11.42452046
14.12501798
3.215186515
6.073187588
5.232639012
2.949429224
5.603792085
8.52462096
6.135766627
8.368898407
8.028701089
9.924303008
13.07584267
7.927364829
11.89456814
11.65486625
13.67657965

```
]
```
Equation for the face drilling – human energy

$$yyy(1,ii) = 0.000319 * power(xx1,0.0495) * power(yy2,-0.5893)$$

$$* power(zz3,0.4369) * power(xx4,-0.3944)$$

$$* power(yy5,1.0414) * power(zz6,2.0566);$$

The above output and equation of face drilling – human energy is placed instead of face drilling– time output and equation in MATLAB program of face drilling– time. Various output graphs obtained for face drilling – human energy are shown as below by the ANN program (Figures 7.17 to 7.24).

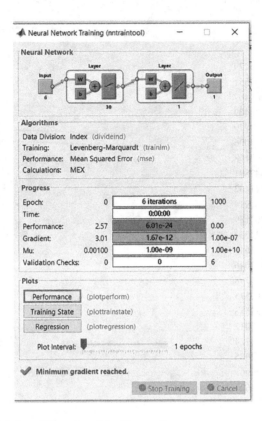

FIGURE 7.17 Training of the network for Human Energy.

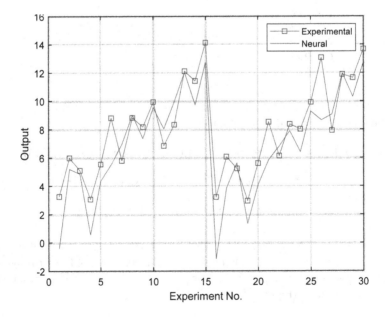

FIGURE 7.18 Graph of comparison with experimental data base and neural prediction for the network for Human Energy.

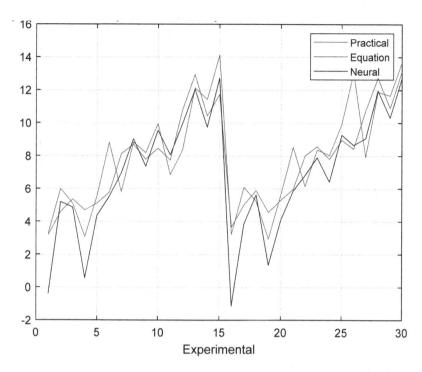

FIGURE 7.19 Graph of comparison with experimental data base, neural network prediction and equation base prediction for the network for Human Energy.

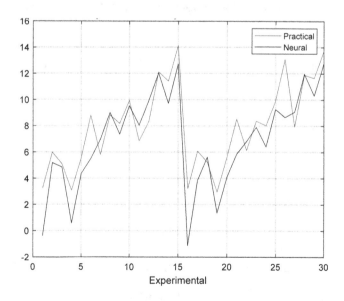

FIGURE 7.20 Graph of comparison with mathematical equation base and neural network prediction for the network for Human Energy.

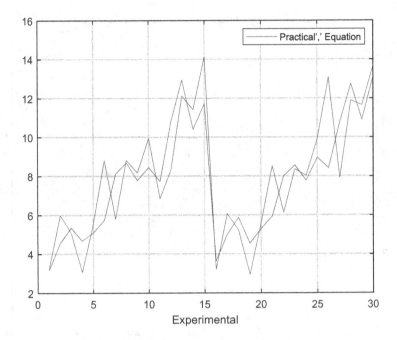

FIGURE 7.21   Graph of comparison with mathematical equation base and experimental data base for the network for Human Energy.

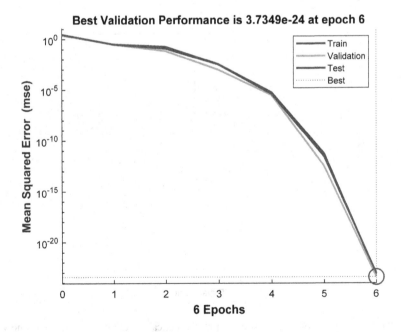

FIGURE 7.22   Graph of training of experimental data for the network for Human Energy.

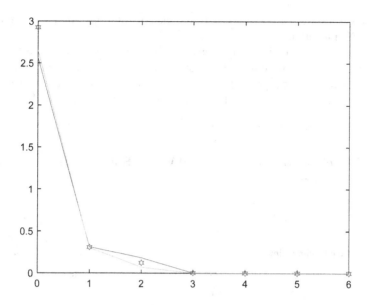

FIGURE 7.23 Graph of training, validation, testing of experimental data for Human Energy.

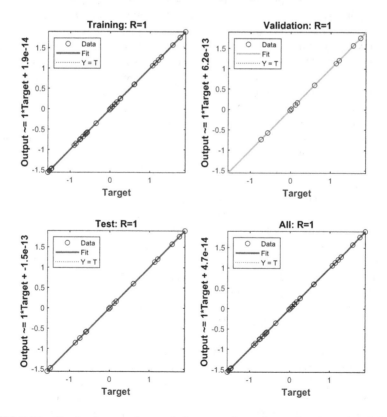

FIGURE 7.24 Graph of regression analysis comparison of actual and computed data by ANN for Human Energy.

These graphs are a representation of the face drilling operation– human energy using ANN simulation.

From the above comparison of phenomenal response by a conventional approach and ANN simulation, it seems to be that the curve obtained by dependent pi terms for face drilling operation time π_{D1}, productivity π_{D2}, and human energy π_{D3} are overlapping due to the less percentage of error, which is on the positive side (Tables 7.1 and 7.2).

TABLE 7.1

Comparison of Experimental Calculated Values, Equation Base Values and ANN-Based Values

Sr. No.	Dependent Variable Face Drilling- Time			Dependent Variable Face Drilling -Productivity			Dependent Variable Face Drilling –Human Energy Consumed		
	Field	Model	ANN	Field	Model	ANN	Field	Model	ANN
1	17.5765	16.2371	15.824	1.8489	1.8581	1.7908	3.2564	3.1699	−0.4022
2	17.6471	19.0256	19.4868	2.0069	1.9593	1.9303	5.9857	4.5828	5.1832
3	17.6647	18.4039	19.0022	2.1464	2.0913	2.1283	5.0920	5.3489	4.8525
4	17.6559	21.0472	17.5528	1.9675	1.8965	1.9486	3.0669	4.6854	0.5555
5	17.6059	16.9429	16.8701	2.1594	2.1369	2.1222	5.5399	5.1002	4.3504
6	30.1818	28.4050	30.1854	1.7945	1.8097	1.7646	8.8063	5.7503	5.5057
7	30.3773	33.7357	33.0565	1.9120	1.8840	1.8942	5.8010	8.1189	6.9570
8	30.3545	28.5315	33.6148	2.0178	2.0823	2.0188	8.8228	8.6897	9.0200
9	30.3136	32.6826	29.5903	1.9431	1.9248	1.9246	8.1745	7.7840	7.3570
10	30.2152	31.2331	29.6781	2.0339	2.0232	2.0252	9.9431	8.4498	9.5405
11	42.2545	37.9251	36.7611	1.7945	1.8246	1.7934	6.8513	7.7371	8.0485
12	42.4909	43.4007	41.1782	1.9003	1.9125	1.9192	8.3391	10.8011	9.9445
13	42.4455	48.6529	42.3912	1.9927	1.9431	2.0094	12.1251	12.9398	12.0469
14	42.5000	40.4358	41.4876	1.9546	1.9342	1.9249	11.4245	10.4173	9.7285
15	42.2561	41.5840	40.9297	2.0583	1.9994	1.9902	14.1250	11.7259	12.7323
16	17.5765	16.3049	16.0624	1.8489	1.9124	1.7709	3.2152	3.6483	−1.1361
17	17.6471	18.9054	18.2318	2.0069	1.9925	1.9184	6.0732	5.0088	3.8515
18	17.6647	18.4034	19.994	2.1464	2.1163	2.1056	5.2326	5.8793	5.6204
19	17.6559	20.0667	14.8369	1.9675	1.9308	1.9357	2.9494	4.5545	1.3508
20	17.6059	16.2004	15.3595	2.1594	2.1942	2.1082	5.6038	5.3012	4.0793
21	30.1818	27.1202	28.2221	1.7945	1.8590	1.7860	8.5246	5.9543	5.8499
22	30.3773	32.0032	29.9827	1.9120	1.9206	1.8904	6.1358	8.0037	6.8145
23	30.3545	27.0104	29.9815	2.0178	2.0943	2.0130	8.3689	8.5659	7.8965
24	30.3136	31.2243	27.7058	1.9431	1.9261	1.9335	8.0287	7.7902	6.4122
25	30.2152	29.9133	28.9511	2.0339	2.0537	2.0209	9.9243	8.9600	9.2689
26	42.2545	36.4652	39.1127	1.7945	1.8296	1.7757	13.0758	8.4191	8.6489
27	42.4909	40.8348	41.0862	1.9003	1.9117	1.8933	7.9274	10.7602	9.0585
28	42.4455	45.7607	40.7326	1.9927	1.9365	1.9778	11.8946	12.7444	11.9700
29	42.5000	40.1657	39.0914	1.9546	1.9890	1.9259	11.6549	10.9021	10.3110
30	42.2561	41.6955	40.6512	2.0583	2.0927	2.0001	13.6766	13.1328	12.7340

TABLE 7.2

Comparison between Observed and Computed Values of Dependent pi Term for Face Drilling Operation

Dependent Pi Term	π_{D2} Time	π_{D3} Productivity	π_{D1} Energy
Mean Field	30.1026	1.9687	7.1880
Mean ANN	29.2537	1.9413	6.9384
Mean Model	30.0106	1.9680	7.809
Mean absolute error performance function	1.6512	0.0298	1.4782
Mean-squared error performance function	4.1554	0.0015	3.5699

8 Sensitivity Analysis

Pramod Belkhode
Laxminarayan Institute of Technology

J. P. Modak
Visvesvaraya National Institute of Technology and
JD College of Engineering and Management

V. Vidyasagar
Power Systems Training Institute

P. B. Maheshwary
JD College of Engineering and Management

CONTENTS

8.1 SENSITIVITY ANALYSIS OF FACE DRILLING OPERATION

Influences of various independent π terms have been studied by analyzing the indices of various independent pi terms in the models. Through the technique of sensitivity analysis, the change in the value of a dependent π term caused due to an introduced change in the value of individual independent π term is evaluated. In this case, the change of $\pm 10\%$ is introduced in the individual π terms independently (one at a time). Thus, the total range of the introduced change is 20%. The effect of this introduced change on the percentage change value of the dependent π term is evaluated. The average value of the change in the dependent π terms due to the introduced change of 20% in each π term is shown in Tables 8.1 and 8.2. These values are plotted on the graph shown in Figures 8.1–8.3.

The sensitivity as evaluated is represented and discussed below.

TABLE 8.1

Sensitivity Analysis for Time of Operation (Z1), Productivity (Z2), and Human Energy (Z3) in Face Drilling Operation

Sensitivity Analysis for Dependent π term

	π 1	π 2	π 3	π 4	π 5	π 6	Z1	Z2	Z3
Avg.	**0.0657**	0.00023	6.54610	1.08698	24.96700	1.67757	30.10056	1.90417	7.15566
0.10	**0.0723**	0.00023	6.54610	1.08698	24.96700	1.67757	30.15598	1.90316	7.18950
−0.10	**0.0591**	0.00023	6.54610	1.08698	24.96700	1.67757	30.03942	1.90530	7.11844
				% Change			0.38726	−0.11238	0.99308
Avg.	0.0657	**0.00023**	6.54610	1.08698	24.96700	1.67757	30.10056	1.90417	7.15566
0.10	0.0657	**0.00025**	6.54610	1.08698	24.96700	1.67757	30.19021	1.86756	6.76483
−0.10	0.0657	**0.00020**	6.54610	1.08698	24.96700	1.67757	30.00178	1.94548	7.61403
				% Change			0.62600	−4.09213	−11.86751
Avg.	0.0657	0.00023	**6.54610**	1.08698	24.96700	1.67757	30.10056	1.90417	7.15566
0.10	0.0657	0.00023	**7.20071**	1.08698	24.96700	1.67757	33.27766	1.85518	7.45992
−0.10	0.0657	0.00023	**5.89149**	1.08698	24.96700	1.67757	26.94022	1.95984	6.83373
				% Change			21.05424	−5.49658	8.75088
Avg.	0.0657	0.00023	6.54610	**1.08698**	24.96700	1.67757	30.10056	1.90417	7.15566
0.10	0.0657	0.00023	6.54610	**1.19568**	24.96700	1.67757	27.79576	1.94545	6.89167
−0.10	0.0657	0.00023	6.54610	**0.97828**	24.96700	1.67757	32.87144	1.85957	7.45927
				% Change			−16.86240	4.51037	−7.93222
Avg.	0.0657	0.00023	6.54610	1.08698	**24.96700**	1.67757	30.10056	1.90417	7.15566
0.10	0.0657	0.00023	6.54610	1.08698	**27.46370**	1.67757	32.11510	1.95816	7.90234
−0.10	0.0657	0.00023	6.54610	1.08698	**22.47030**	1.67757	28.02033	1.84623	6.41206
				% Change			13.60362	5.87785	20.82662
Avg.	0.0657	0.00023	6.54610	1.08698	24.96700	**1.67757**	30.10056	1.90417	7.15566
0.10	0.0657	0.00023	6.54610	1.08698	24.96700	**1.84533**	30.83198	1.98942	8.70518
−0.10	0.0657	0.00023	6.54610	1.08698	24.96700	**1.50981**	29.31219	1.81418	5.76162
				% Change			5.04904	9.20281	41.13610

8.1.1 EFFECT OF INTRODUCED CHANGE ON THE DEPENDENT π TERM—TIME OF FACE DRILLING OPERATION

In this model, when a total change of 20% is introduced in the value of independent π term π_1, a change of +0.3872% occurs in the value of Z_1, that is, time of face drilling (computed from the model). The change brought in the value of Z_1 because of the change in the value of the other independent π terms namely π_2 is +0.6262%. Similarly, the change of about +21.0542%, −16.8624%, +13.6036%, and +5.0490% takes place in the value of Z_1 because of the change in the values of π_3, π_4, π_5, and π_6, respectively.

It can be observed that the highest change takes place in Z_1 because of the π term π_3, whereas the least change takes place due to π_1. Thus, π_3 is most sensitive π term and π_1 term is the least sensitive π term. The sequence of various π terms in the descending order of sensitivity is π_5, π_6, π_2, and π_4.

TABLE 8.2

Percentage Change in Time (Z1), Productivity (Z2), and Human Energy (Z3) in Face Drilling Operation with ±10% Variation of Mean Value of Each π Term

Sensitivity Chart

π Terms	Z1	Z2	Z3
π 1	0.3872	−0.1123	0.9930
π 2	0.6260	−4.0921	−11.8675
π 3	21.0542	−5.4965	8.7508
π 4	−16.8624	4.5103	−7.9322
π 5	13.6036	5.8778	20.8266
π 6	5.0490	9.2028	41.1361

8.1.2 EFFECT OF INTRODUCED CHANGE ON THE DEPENDENT π TERM–PRODUCTIVITY OF FACE DRILLING OPERATION

In this model, when a total change of 20% is introduced in the value of independent π term π_1 a change of −0.1123% occurs in the value of Z_2, that is, productivity of face drilling (computed from the model). The change brought in the value of Z_2 because of the change in the value of the other independent π term π_2 is−4.0921%. Similarly, the change of about −5.4965%, +4.5103%, +5.8778%, and +9.2028% takes place in the value of Z_2 because of the change in the values of π_3, π_4, π_5, and π_6, respectively.

It can be observed that the highest change takes place in Z_2 because of the π term π_6, whereas the least change takes place due to π_4. Thus, π_6 is most sensitive π term and π_3 term is the least sensitive π term. The sequence of the various π terms in the descending order of sensitivity is π_5, π_1, π_2, and π_3.

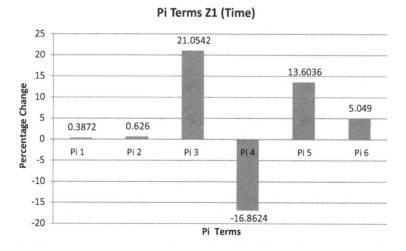

FIGURE 8.1 Percentage change in π terms variation for dependent term Z_1 (Time).

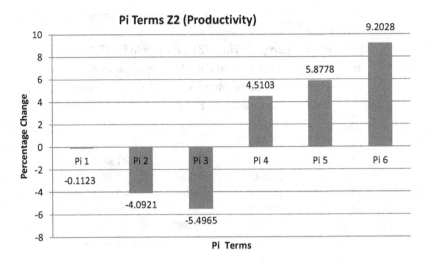

FIGURE 8.2   Percentage change in π terms variation for dependent term Z_2 (Productivity).

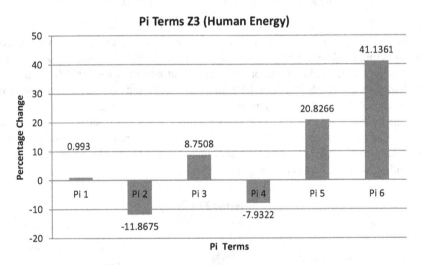

FIGURE 8.3   Percentage change in π terms variation for dependent term Z_3 (Human Energy).

8.1.3   EFFECT OF INTRODUCED CHANGE ON THE DEPENDENT π TERM—HUMAN ENERGY CONSUMED IN FACE DRILLING OPERATION

In this model, when a total change of 20% is introduced in the value of independent π term π_1, a change of +0.9930% occurs in the value of Z_3, that is, human energy consumed in face drilling (computed from the model). The change brought in the value of Z_3 because of the change in the value of the other independent π term π_2 is −11.8675%. Similarly, the change of about +8.7508%, −7.9322%, +20.8266%, and +41.1361% takes place in the value Z_3 because of the change in the values of π_3, π_4, π_5, and π_6, respectively.

It can be observed that the highest change takes place in Z_3 because of the π term π_6, whereas the least change takes place due to π_1. Thus, π_6 is most sensitive π term and π_1 term is the least sensitive π term. The sequence of the various π terms in the descending order of sensitivity is π_5, π_3, π_4, and π_2.

8.2   OPTIMIZATION OF MODELS

The models have been developed for the phenomenon of face drilling in mines. The ultimate objective of this work is not merely developing the models but to finding out the best set of independent variables, which will result in maximization/mini-mization of the objective function [32]. In this case, we have three different models corresponding to the time of face drilling (Z_1), productivity of face drilling (Z_2), and human energy consumed in face drilling operation (Z_3). Three objective functions will correspond to these models. The model for the productivity of face drilling (Z_2) needs to be maximized, whereas models for time of face drilling (Z_1) and human energy required in face drilling (Z_3) need to be minimized. The models are in non-linear form; hence, they are to be converted into a linear form for optimization pur-pose. This is achieved by taking the log on both sides of the model. To maximize the linear function, we can use the linear programming technique as shown below:

For the dependent p term (Z_1), we have

$$(Z_1) = K_1 * \left[(\pi_1)^{a1} * (\pi_2)^{b1} * (\pi_3)^{c1} * (\pi_4)^{d1} * (\pi_5)^{e1} * (\pi_6)^{f1} \right] \qquad (8.1)$$

Taking log on both sides of the equation, we have

$$Log(Z_1) = Log\ K1 + a1 * Log\ (\pi_1) + b1 * Log(\pi_2) + c1 * Log(\pi_3) +$$

$$d1 * Log(\pi_4) + e1 * Log(\pi_5) + f1 * Log(\pi_6)$$

Let, $Log(Z_1) = Z$, $Log\ K1 = K1'$, $log(\pi_1) = X1$, $log(\pi_2) = X2$, $log(\pi_3) = X3$, $log(\pi_4) = X4$, $log(\pi_5) = X5$, and $log(\pi_6) = X6$, then the linear model in the form of first degree polynomial can be written as follows:

$$Z = K1' + a_1 * X1 + b_1 * X2 + c_1 * X3 + d_1 * X4 + e_1 * X5 + f_1 * X6 \qquad (8.2)$$

Thus, the equation will be the objective function for the optimization or to be very specific for maximization for formulation of the linear programming problem. The next task is to define the constraints for the problem. The constraints can be the boundaries defined for various independent p terms involved in the function. During the field study, the ranges for each independent π term have been observed. These ranges will be the constraints for the problem. Thus, there will be two constraints for each independent variable as shown below.

If we denote the maximum and minimum values of a dependent π term Z_1 by π_1 max. and π_1 min. by then, the first two constraints for the problem will be obtained by taking log of these quantities and by substituting the values of multipliers of all other variables except the one under consideration equal to Zero. Let the log of the limits be defined, as C1 and C2 {i.e., C1 = $\log(\pi_1$ max.) and C2 = $\log(\pi_1$ min.)}. Thus, the equations of the constraints will be as follows:

$$1 * X1 + 0 * X2 + 0 * X3 + 0 * X4 + 0 * X5 + 0 * X6 \leq C1$$

$$1 * X1 + 0 * X2 + 0 * X3 + 0 * X4 + 0 * X5 + 0 * X6 \geq C2$$

The other constraints can be likewise found as follows:

$$0 * X1 + 1 * X2 + 0 * X3 + 0 * X4 + 0 * X5 + 0 * X6 \leq C3$$

$$0 * X1 + 1 * X2 + 0 * X3 + 0 * X4 + 0 * X5 + 0 * X6 \geq C4$$

$$0 * X1 + 0 * X2 + 1 * X3 + 0 * X4 + 0 * X5 + 0 * X6 \leq C5$$

$$0 * X1 + 0 * X2 + 1 * X3 + 0 * X4 + 0 * X5 + 0 * X6 \geq C6$$

$$0 * X1 + 0 * X2 + 0 * X3 + 1 * X4 + 0 * X5 + 0 * X6 \leq C7$$

(8.3)

$$0 * X1 + 0 * X2 + 0 * X3 + 1 * X4 + 0 * X5 + 0 * X6 \geq C8$$

$$0 * X1 + 0 * X2 + 0 * X3 + 0 * X4 + 1 * X5 + 0 * X6 \leq C9$$

$$0 * X1 + 0 * X2 + 0 * X3 + 0 * X4 + 1 * X5 + 0 * X6 \geq C10$$

$$0 * X1 + 0 * X2 + 0 * X3 + 0 * X4 + 0 * X5 + 1 * X6 \leq C11$$

$$0 * X1 + 0 * X2 + 0 * X3 + 0 * X4 + 0 * X5 + 1 * X6 \geq C12$$

After solving this linear programming problem, one get the minimum value of the Z and the set of values of the variables to achieve this minimum value. The values of the independent pi terms can then be obtained by finding the antilog of the values of Z, X1, X2, X3, X4, X5, and X6. The actual values of the multipliers and the variables are found and substituted in Equations 8.3, and the actual problem in this case can be stated as below. This can now be solved as a linear programming problem using MS Solver available in MS Excel. Thus, the actual problem is to maximize Z, where

$$Z = K1' + a_1 * X1 + b_1 * X2 + c_1 * X3 + d_1 * X4 + e_1 * X5 + f_1 * X6 \qquad (8.4)$$

$$Z = \log(0.6025) + 0.0193 * \log(\pi_1) + 0.0312 * \log(\pi_2) + 1.0528 * \log(\pi_3)$$

$$- 0.8358 * \log(\pi_4) + 0.6797 * \log(\pi_5) + 0.2519 * \log(\pi_6)$$

$$Z = \log(0.6025) + (0.0193 * X1) + (0.0312 * X2) + (1.0528 * X3)$$

$$+ (-0.8358 * X4) + (0.6797 * X5) + (0.2519 * X6)$$

Subject to following constraints:

$$1 * X1 + 0 * X2 + 0 * X3 + 0 * X4 + 0 * X5 + 0 * X6 \leq -0.8173$$

$$1 * X1 + 0 * X2 + 0 * X3 + 0 * X4 + 0 * X5 + 0 * X6 \geq -1.5724$$

The other constraints can be likewise found as below:

$$0 * X1 + 1 * X2 + 0 * X3 + 0 * X4 + 0 * X5 + 0 * X6 \leq -3.2863$$

$$0 * X1 + 1 * X2 + 0 * X3 + 0 * X4 + 0 * X5 + 0 * X6 \geq -4.0065$$

$$0 * X1 + 0 * X2 + 1 * X3 + 0 * X4 + 0 * X5 + 0 * X6 \leq 1.0320$$

$$0 * X1 + 0 * X2 + 1 * X3 + 0 * X4 + 0 * X5 + 0 * X6 \geq 0.4684$$

$$0 * X1 + 0 * X2 + 0 * X3 + 1 * X4 + 0 * X5 + 0 * X6 \leq 0.1842$$

$$0 * X1 + 0 * X2 + 0 * X3 + 1 * X4 + 0 * X5 + 0 * X6 \geq -0.0858$$
(8.5)
$$0 * X1 + 0 * X2 + 0 * X3 + 0 * X4 + 1 * X5 + 0 * X6 \leq 1.4308$$

$$0 * X1 + 0 * X2 + 0 * X3 + 0 * X4 + 1 * X5 + 0 * X6 \geq 1.3708$$

$$0 * X1 + 0 * X2 + 0 * X3 + 0 * X4 + 0 * X5 + 1 * X6 \leq 0.2655$$

$$0 * X1 + 0 * X2 + 0 * X3 + 0 * X4 + 0 * X5 + 1 * X6 \geq 0.1906$$

On solving the above problem with the MS solver, we get,

$$X1 = -1.5724, X2 = -4.0065, X3 = 0.4684, X4 = 0.1842,$$

$$X5 = 1.3708, X6 = 0.1906 \text{ and } Z = 0.9435$$

Thus, Z1 Min. = Antilog(0.9435) = 8.7814 and corresponding to this, the values of the Z1 Min. the values of independent π terms are obtained by taking the antilog of X1, X2, X3, X4, X5, X6, and X7. These values are 0.0267, 0.0000985, 2.9403, 1.5285, 23.49, and 1.5510, respectively.

Similar procedure has been adopted to optimize the models for Z2 and Z3.

Z2 Max. = Antilog(0.5120) = 3.2515 and corresponding to this value of Z2 Max., the values of independent pi terms are obtained by taking antilog of X1, X2, X3, X4, X5, and X6. These values are 0.0267, 0.0000985, 2.9403, 1.5285, 26.97, and 1.8431, respectively.

Z3 Min. = Antilog(0.3148) = 2.0647 and corresponding to this value of Z3 Min., the values of the independent π terms are obtained by taking the antilog of 0.0267, 0.000517, 2.9403, 1.5285, 23.49, and 1.5510, respectively.

9 Analysis of Performance of the Model

Pramod Belkhode
Laxminarayan Institute of Technology

J. P. Modak
Visvesvaraya National Institute of Technology and
JD College of Engineering and Management

V. Vidyasagar
Power Systems Training Institute

Pratibha Agrawal
Laxminarayan Institute of Technology

CONTENTS

9.1 DEVELOPED MODELS FOR DEPENDENT VARIABLES

The exact forms of models obtained for the dependent variables time of operation, productivity, and human energy consumed in face drilling operation are as follows:

$$(Z_1) = 0.6025 * (\pi_1)^{0.0193} * (\pi_2)^{0.0312} * (\pi_3)^{1.0528} * (\pi_4)^{-0.8358} * (\pi_5)^{0.6797} * (\pi_6)^{0.2519} \quad (9.1)$$

$$(Z_2) = 0.1706 * (\pi_1)^{-0.0056} * (\pi_2)^{-0.2037} * (\pi_3)^{-0.2735} * (\pi_4)^{0.2250} * (\pi_5)^{0.2933} * (\pi_6)^{0.4595} \quad (9.2)$$

$$(Z_3) = 0.000319 * (\pi_1)^{0.0495} * (\pi_2)^{-0.5893} * (\pi_3)^{0.4369} * (\pi_4)^{-0.3944} * (\pi_5)^{1.0414} * (\pi_6)^{2.0566} \quad (9.3)$$

In the above equations, (Z_1), (Z_2), and (Z_3) are relating to response variable time of face drilling activity, (Z_2) is relating to response variable productivity of face

95

drilling, and (Z_3) is relating to response variable human energy consumed in the activity, respectively.

Interpretation of the Model
Interpretation of model is being reported in terms of several aspects, namely, viz (i) interpretation of curve-fitting constant K, (ii) order of influence of various inputs (causes) on outputs (effects), and (iii) relative influence of causes on effect.

a. Interpretation of Curve-Fitting Constant (K_1) for Z_1
The value of curve-fitting constant in this model for (Z_1) is 0.6025. This collectively represents the combined effect of all extraneous variables such as lower aerobic capacity of miners, physiological and biomechanical demands of doing manual work in vertical space restrictions, and so on. Further, as it is positive, this indicates that these causes have increasing influence on the time of face drilling activity (Z_1). In addition, its value is even less than one. The estimated causes are in excess of real magnitudes of causes. Hence, it is in this respect that there is a scope to estimate the causes with more precision while executing another fresh attempt to refine the models.

Analysis of the Model for Dependent π Term Z_1
1. The absolute index of π_3 is the highest, that is, 1.0528. Thus, the term related to the specification of drill machine and process parameters are the most influencing π term in this model. The value of this index is positive indicating that the π term related to time of face drilling activity, that is, (Z_1) is directly proportional this π term. The specifications of drill machine and process parameters, that is, π_3, indicate that more weight of drill rod, weight of compressed air hose, weight of jack hammer, shear strength and density of ore with low compression air pressure will increase more time required for face drilling. This suggests that high process parameters such as increase in penetration rate, compressed air pressure with less weight of drill machine will reduce the time required for the face drilling.
2. The absolute index of π_1 is the lowest, that is, 0.0193. Thus, the term related to anthropometric data of the miner is the least influencing π term in this model. The value of this index is positive indicating that the π term related to time of face drilling activity (Z_1) is directly proportional to the term related to anthropometric data of the miner at the workstation. As the age, enthusiasm, general health status of miner related to $[\pi_1]$ deteriorates thus increases the time for face drilling.

    Good planning at workstation regarding the distribution of work as per the age, general health status, and motivation of the miner about their work would reduce the time required for the face drilling result in enhancing the productivity.
3. The sequence of influence on Z1 of other independent π terms of this model, that is, π_4, π_5, π_6, and π_2, respectively, having absolute indices as −0.8358, 0.6797, 0.2519, and 0.0312 in the reducing order.

The time of face drilling activity (Z_1) is inversely proportional to the term related to the speed and penetration rate $[\pi_4]$ with the index as -0.8358. With the increase in drilling speed, and penetration rate, time required for face drilling reduces.

Similarly, the time of face drilling activity (Z_1) is directly proportional to the term related to the temperature and relative humidity at the workstation $[\pi_5]$ with the index as 0.6797, increasing the time required for face drilling as the miner gets exhausted quickly with the increase in ambient temperature and relative humidity at the workstation.

The time of face drilling activity (Z_1) is directly proportional to the term related to illumination $[\pi_6]$ with the index as 0.2519 indicates that in the good illuminated workstation, less time is required for face drilling suggesting that artificial lighting (through petromax light/battery operated lights/LED instead of conventional bulbs) will reduce the time required for the face drilling.

Similarly, the time of face drilling activity (Z_1) is directly proportional to the term related to the specification of drill rod $[\pi_2]$ with parameters as [Lr/ Dr], [Dc/ Dr], and [Lc/Dr] esp. as the length of drill rod{Lr} increases the time required for the face drilling increases with the index as 0.0312.

b. Interpretation of Curve-Fitting Constant (K_2) of π Term Z_2

The value of curve-fitting constant in this model for (Z_2) is 0.1706. This collectively represents the combined effect of all extraneous variables such as lower aerobic capacity of miners, physiological and biomechanical demands of doing manual work in vertical space restrictions, and so on. Further, as it is positive, this indicates that these causes have increasing influence on the productivity of face drilling (Z_2).

Analysis of the Model for Dependent π Term Z_2

1. The absolute index of π_6 is the highest, namely, 0.4595. Thus, the productivity of face drilling (Z_2) is directly proportional to the term related to the illumination $[\pi_7]$ with the index as 0.4595. With the increase in illumination [I], the productivity of face drilling increases.

2. The absolute index of π_1 is the lowest, that is, 0.0056. The productivity of face drilling (Z_2) is inversely proportional to the term related to the anthropometric data of the miner $[\pi_1]$ with the index as -0.0056. With the suitable age of the miner and good habits, the miner makes use of his anthropometric dimensions to the advantage of increasing productivity of face drilling.

3. The sequence of influence of other independent π terms present on this model is π_5, π_4, π_3, and π_2 having absolute indices as 0.2933, 0.2250, 0.2735, and 0.2037, respectively.

The index of π_5 is positive indicating that the productivity of face drilling (Z_2) varies directly as to the term related to ambient temperature [θ] and relative humidity at the workstation $[\pi_5]$ with the index as 0.2933, increasing the productivity of face drilling with good ambient temperature and relative humidity during the face drilling.

Similarly, the productivity of face drilling activity (Z_2) is directly proportional to the term related to the speed and penetration rate of drill machine [π_4] with the index as 0.225, as the speed of drilling machine [N] increases the productivity of face drilling increases.

The productivity of face drilling activity (Z_2) is inversely proportional to the term related to Specifications of drill machine/process parameters [π_3] is the most influencing π term in this model. The value of this index is negative indicating that the productivity of face drilling (Z_2) is inversely proportional to the term related to specifications of drill machine/process parameters [π_3] with the index as -0.2735. As the values of weight of jack hammer (Wj), shear strength of ore (So), shear strength of mica (Ss), hardness of drill rod (Hr), density of ore (Do), and density of mica (Ds) increase the productivity of face drilling falls.

The productivity of face drilling activity (Z_2) is inversely proportional to the term related to the specifications of drill rod [p_2] with parameters as [Lr/Dr], [Dc/Dr], [Lc/Dr] esp. as the length of drill rod [Lr] increases the productivity of face drilling decreases with the index as -0.2037.

c. **Interpretation of Curve-Fitting Constant (K_3) for Z_3**

The value of curve-fitting constant in this model for (Z_3) is 0.000319. This collectively represents the combined effect of all extraneous variables such as lower aerobic capacity of miners, physiological and biomechanical demands of doing manual work in vertical space restrictions, and so on. Further, it is positive, indicating that these causes have increasing influence on the human energy consumed [Z_3]. This indicates that more precise accounting of causes of face drilling operation is essential in future attempts of formulating the same modeling.

Analysis of the Model for Dependent π Term Z_3

1. The absolute index of π_6 is the highest, that is, 2.0566. Thus, the term related to illumination at the workstation [π_6] is the most influencing π term in this model. The value of this index is positive indicating that the human energy consumed (Z_3) is directly proportional to the term related to illumination at the workstation [π_6] with the index as 2.0566, as human energy is properly applied with illumination at the workstation. Indirectly, it shows the estimation of presently considered causes in magnitude with more precision.

2. The absolute index of π_1 is the lowest, that is, 0.0495. Thus, the term related to anthropometric data of miner [π_1] is the least influencing π term in this model. The value of this index is positive indicating that human energy consumed (Z_3) is directly proportional to the anthropometric data of miner [π_1] with its index as 0.0495, indicating that the bent posture enhances the human energy requirements.

3. The sequence of influence of other independent π terms present in this model is π_5, π_2, π_3, and π_4 with absolute indices as 1.0414, 0.5893, 0.4369, and 0.3944, respectively.

The human energy consumed (Z_3) is directly proportional to the term related to the ambient temperature at the workstation [π_5] with the index as 1.0414 indicating that with the increase in ambient temperature [θ] and relative Humidity [ϕ], the human energy requirements increase in face drilling.

The human energy consumed (Z_3) is inversely proportional to the term related to the specifications of drill rod [π_2] with parameters as [Lr/Dr], [Dc/Dr], [Lc/Dr] esp. as the length of Drill rod [Lr] increases the human energy consumption of face drilling increases with the index as −0.5893.

Similarly, the human energy consumed (Z_3) is directly proportional to the term related to the specifications of drilling machine/process parameters [π_3] with the index as 0.4369. As compressed air pressure [Pa] reduces and the values of weight of jack hammer (Wj), shear strength of ore (So), shear strength of mica (Ss), hardness of drill rod (Hr), density of ore (Do), density of mica (Ds) increase human energy requirements of face drilling increases.

The human energy consumed (Z_3) is inversely proportional to the term related to the speed and penetration rate [π_4] with the index as −0.3944. The increase in drilling speed and penetration rate reduces the human energy consumption for face drilling.

9.2 ANALYSIS OF PERFORMANCE OF THE MODELS OF FACE DRILLING OPERATION

The models have been formulated mathematically as well as using the ANN. Now, we have three sets of values three dependent π terms, namely, values computed by field study observations, values computed by mathematical models, and the values obtained by ANN, match very well with each other. This is justified by calculating their respective mean values and standard errors of estimation. The comparison of dependent π terms obtained by field observation-based calculated values, exponential form of model-based values, and ANN-based values are indicated in Table 9.1.

From these comparisons, phenomena response by a conventional approach and ANN simulation, it seems that the curve obtained by dependent π terms for face drilling operation –Time π_{D1}, Productivity π_{D2}, and Human Energy π_{D3} are overlapping due to less percentage error which is on positive side and gives an accurate relationship between ANN simulation and field data (Table 9.2).

From the above tables, it seems that the mathematical models and ANN developed using MATLAB can be successfully used for computation of dependent π terms for a given set of independent π terms.

9.3 RELIABILITY OF THE MODELS

Before taking up the step of sensitivity of inputs, it is necessary to decide the validity of the model. This is so because though, the observed data are purified by known statistics-based techniques, there is a chance of some impure data entering in the mathematical processing of the data in spite of even using MATLAB.

TABLE 9.1

Comparison of Experimental Calculated Values, Equation-Based Values, and ANN-Based Values

Sr. No.	Dependent Variable Face Drilling – Time			Dependent Variable Face Drilling – Productivity			Dependent Variable Face Drilling – Human Energy Consumed		
	Field	Model	ANN	Field	Model	ANN	Field	Model	ANN
1	17.5765	16.2371	15.824	1.8489	1.8581	1.7908	3.2564	3.1699	−0.4022
2	17.6471	19.0256	19.4868	2.0069	1.9593	1.9303	5.9857	4.5828	5.1832
3	17.6647	18.4039	19.0022	2.1464	2.0913	2.1283	5.0920	5.3489	4.8525
4	17.6559	21.0472	17.5528	1.9675	1.8965	1.9486	3.0669	4.6854	0.5555
5	17.6059	16.9429	16.8701	2.1594	2.1369	2.1222	5.5399	5.1002	4.3504
6	30.1818	28.4050	30.1854	1.7945	1.8097	1.7646	8.8063	5.7503	5.5057
7	30.3773	33.7357	33.0565	1.9120	1.8840	1.8942	5.8010	8.1189	6.9570
8	30.3545	28.5315	33.6148	2.0178	2.0823	2.0188	8.8228	8.6897	9.0200
9	30.3136	32.6826	29.5903	1.9431	1.9248	1.9246	8.1745	7.7840	7.3570
10	30.2152	31.2331	29.6781	2.0339	2.0232	2.0252	9.9431	8.4498	9.5405
11	42.2545	37.9251	36.7611	1.7945	1.8246	1.7934	6.8513	7.7371	8.0485
12	42.4909	43.4007	41.1782	1.9003	1.9125	1.9192	8.3391	10.8011	9.9445
13	42.4455	48.6529	42.3912	1.9927	1.9431	2.0094	12.1251	12.9398	12.0469
14	42.5000	40.4358	41.4876	1.9546	1.9342	1.9249	11.4245	10.4173	9.7285
15	42.2561	41.5840	40.9297	2.0583	1.9994	1.9902	14.1250	11.7259	12.7323
16	17.5765	16.3049	16.0624	1.8489	1.9124	1.7709	3.2152	3.6483	−1.1361
17	17.6471	18.9054	18.2318	2.0069	1.9925	1.9184	6.0732	5.0088	3.8515
18	17.6647	18.4034	19.994	2.1464	2.1163	2.1056	5.2326	5.8793	5.6204
19	17.6559	20.0667	14.8369	1.9675	1.9308	1.9357	2.9494	4.5545	1.3508
20	17.6059	16.2004	15.3595	2.1594	2.1942	2.1082	5.6038	5.3012	4.0793
21	30.1818	27.1202	28.2221	1.7945	1.8590	1.7860	8.5246	5.9543	5.8499
22	30.3773	32.0032	29.9827	1.9120	1.9206	1.8904	6.1358	8.0037	6.8145
23	30.3545	27.0104	29.9815	2.0178	2.0943	2.0130	8.3689	8.5659	7.8965
24	30.3136	31.2243	27.7058	1.9431	1.9261	1.9335	8.0287	7.7902	6.4122
25	30.2152	29.9133	28.9511	2.0339	2.0537	2.0209	9.9243	8.9600	9.2689
26	42.2545	36.4652	39.1127	1.7945	1.8296	1.7757	13.0758	8.4191	8.6489
27	42.4909	40.8348	41.0862	1.9003	1.9117	1.8933	7.9274	10.7602	9.0585
28	42.4455	45.7607	40.7326	1.9927	1.9365	1.9778	11.8946	12.7444	11.9700
29	42.5000	40.1657	39.0914	1.9546	1.9890	1.9259	11.6549	10.9021	10.3110
30	42.2561	41.6955	40.6512	2.0583	2.0927	2.0001	13.6766	13.1328	12.7340

The approach to decide the validity would be to substitute in the model known inputs for every observation and decide the difference in response by model and actually observed response. This will give pattern of distribution of error and frequency of its occurrence. Using this distribution and literature on reliability would establish the reliability of the model (Tables 9.2–9.5).

TABLE 9.2

Comparison between Observed and Computed Values of Dependent pi Term for Face Drilling Operation

Dependent Pi Term	π_{D2} Time	π_{D3} Productivity	π_{D1} Energy
Mean Field	30.1026	1.9687	7.1880
Mean ANN	29.2537	1.9413	6.9384
Mean Model	30.0106	1.9680	7.809
Mean absolute error performance function	1.6512	0.0298	1.4782
Mean squared error performance function	4.1554	0.0015	3.5699

TABLE 9.3

Reliability of Model Developed for Dependent Variable: Face Drilling Time

%Error (fi)	Frequency (xi)	fi*xi
0	1	0
1	2	2
2	1	2
3	4	12
4	3	12
5	3	15
6	1	6
7	7	49
10	2	20
11	2	22
13	2	26
14	1	14
19	1	19
Σ95	Σ30	Σ199
Mean Error	$\sum fi * xi/xi$	6.63333333
Reliability	(100-Mean Error)	93.3666667

Note: Reliability of model for face drilling time is 93.36%.

TABLE 9.4

Reliability of Model Developed for Dependent Variable: Face Drilling Productivity

%Error (fi)	Frequency (xi)	fi*xi
1	1	1
2	3	6

(Continued)

TABLE 9.4 (*Continued*)
**Reliability of Model Developed for Dependent
Variable: Face Drilling Productivity**

%Error (fi)	Frequency (xi)	fi*xi
3	1	3
4	1	4
5	2	10
6	2	12
7	2	14
8	2	16
9	1	9
12	2	24
13	1	13
15	1	15
16	1	16
17	1	17
23	1	23
29	1	29
30	1	30
34	1	34
35	2	70
39	1	39
52	1	52
54	1	54
Σ414	Σ30	Σ491
Mean Error	$\sum fi * xi/xi$	16.3666667
Reliability	(100-Mean Error)	83.6333333

Note: Reliability of model of face drilling productivity is 83.63%.

TABLE 9.5
**Reliability of Model Developed for Dependent
Variable: Human Energy for Face Drilling**

%Error (fi)	Frequency (xi)	fi*xi
0	10	0
1	10	10
2	5	10
3	5	15
Σ6	Σ30	Σ35
Mean Error	$\sum fi * xi/xi$	1.16666667
Reliability	(100-Mean Error)	98.8333333

Note: Reliability of model for human energy consumption in
face drilling is 98.83%.

10 Quantitative Analysis of Face Drilling Operation

Pramod Belkhode
Laxminarayan Institute of Technology

J. P. Modak
Visvesvaraya National Institute of Technology and
JD College of Engineering and Management

V. Vidyasagar
Power Systems Training Institute

P. B. Maheshwary
JD College of Engineering and Management

CONTENT

10.1 QUANTITATIVE STUDIES ON FACE DRILLING OPERATION

The postural discomfort experienced by miners while performing face drilling became the cornerstone for this work. They are not aware as to what extent ergonomic intervention can alleviate their drudgery. Second, the relationship between various inputs such as anthropometry of miners, specifications of drill machine, specification of tools, surrounding environmental conditions and their responses such as time to complete drill, human energy, and productivity of face drilling activity is not known to them quantitatively.

From the quantitative studies on face drilling operation, the following conclusions appear to be justified:

The data have been collected by performing actual field observation. Owing to this, the findings of the present study seem to be reliable.

The exponential form of mathematical model and ANN developed for the phenomenon represents the degree of interaction of various independent variables. This is made possible only by the approach adopted in the investigation. The standard error of estimate of the predicted/computed values of the dependent variables is found to be very low. This gives authenticity to the developed mathematical models and ANN.

The trends for the behavior of the models demonstrated by graphical analysis and sensitivity analysis are found complementary to each other. These trends are found to be truly justified through some possible qualitative ergonomics-based analysis of phenomenon.

Analysis of mathematical models showed that the influence of speed and penetration rate of drill machine on the time of face drilling is significant and increases with poor speed and penetration rate. The response variable "Human energy consumed" is significant and decreases with improvement in speed and penetration rate. The response variable "Productivity" increases moderately with speed and penetration rate.

From the study on the model of the time of face drilling, it is found that the influence of the specification of drilling machine, ambient temperature and relative humidity, illumination are predominant over anthropometry of miner and specifications of tools.

From the study on the model of the productivity of face drilling, it is found that the influence of illumination, speed of drilling machine and penetration rate, relative humidity, and ambient temperature are predominant over anthropometry of miner, process parameters, and specifications of tools.

From the study on the model of the human energy consumed in face drilling, it is found that the influences of the illumination, speed of drilling machine and penetration rate and ambient temperature and relative humidity are predominant effects over process parameters and anthropometric data of miner.

Thus, from these models, "Intensity of interaction of inputs on deciding Response" can be predicted, which will help control the variable for the desired results.

References

1. Collier, S.G., Chan, W.L., Mason, S., and Pethick, A.J. 1986. *Ergonomic Design Handbook for Continuous Miners*. 128 pp., Institute of Occupational Medicine, Edinburgh, Scotland.
2. Schenck, H. Jr., 1967. *Theories of Engineering Experimentation*, First Edition, McGraw Hill Inc., New York
3. Ayoub, M.M., Smith, J.L., Selan, J.L., Chen, H.C., Lee, Y.H., Kim, H.K., and Fernandez, J.E. 1987. Manual materials handling in unusual postures. Technical Report, Department of Industrial Engineering, Texas Tech University, Lubbock, TX.
4. Bobick, T.G. 1987. Analyses of materials-handling systems in underground low-coal mines, in *Proceedings of Bureau of Mines Technology Transfer Seminar, Bureau of Mines Information Circular 9145*, pp. 13–20, U.S. Bureau of Mines, Pittsburgh, PA.
5. Bobick, T.G., Gallagher, S., and Unger, R.L. 1989. Effects of random whole-body vibration on back strength and back endurance, in *Advances in Industrial Ergonomics and Safety*, Ed. A. Mital, pp. 537–544. Taylor & Francis, London.
6. Conway, K. and Unger, R.L. 1991. Ergonomic guidelines for designing and maintaining underground coal mining equipment, in *Workspace Equipment and Tool Design*, Ed. A. Mital and W. Karwowski, pp. 279–302. Elsevier Sciences Publishing Co., Amsterdam.
7. Cornelius, K.M., Redfern, M.S., and Steiner, L.J. 1994. Postural stability after whole-body vibration exposure. *Int. J. Ind. Ergon.* 13(4): 343–352.
8. Gallagher, S. 1991. Acceptable weights and physiological costs of performing combined manual handling tasks in restricted postures. *Ergonomics.* 34(7): 939–952.
9. Helander, M., Conway, E.J., Elliott, W., and Curtin, R. 1980. *Standardization of Controls for Roof Bolter Machines. Phase I. Human Factors Engineering Analysis* (contract HO292007, Canyon Research Group Inc.). Bureau of Mines PFR 170-82, NTIS PB 83-119149, 192 pp.
10. Humphreys, P.W. and Lind, A.R. 1962. The energy expenditure of coal-miners at work. *Br. J. Ind. Med.* 19: 264–275.
11. Klein, B.P., Jensen, R.C., and Sanderson, L.M. 1984. Assessment of workers' compensation claims for back strains/sprains. *J. Occ. Med.* 26(6): 443–448.
12. Long, D.A. 1983. Solving the problem of getting on and off large surface mining equipment. Paper in *Bureau of Mines Information Circular 8947*, pp. 3–16, U.S. Bureau of Mines, Minneapolis, MN.
13. Love, A.C., Unger, R.L., Bobick, T.G., and Fowkes, R.S. 1992. *A Summary of Current Bureau Research into the Effects of Whole-Body Vibration and Shock on Operators of Underground Mobile Equipment*, U.S. Bureau of Mines Report of Investigations 9439, pp. 1–15, Pittsburgh, PA.
14. MacDonald, E.B., Porter, R., Hibbert, C., and Hart, J. 1984. The relationship between spinal canal diameter and back pain in coal miners. *J. Occ. Med.* 26(1): 23–28.
15. Marras, W.S. and Lavender, S. 1988. *An Analysis of Hand Tool Injuries in the Underground Mining Industries*. U.S. Bureau of Mines Final Report JO348043, 234 pp., The Ohio State University, Columbus, OH.
16. Marras, W.S. and Lavender, S. 1991. The effects of method of use, tool design, and roof height on trunk muscle activities during underground scaling bar use. *Ergonomics.* 34(2): 221–232.
17. Randolph, R.F. and Love, A.C. 1991. Ergonomics in mining: Human factors and organizational research on high-technology mining. *Appl. Occ. Environ. Hyg.* 6(7): 577–580.

18. Sanders, M.S., Peay, J., and Bobick, T.G. 1981. Research on the development of personal protective equipment for underground mines. Paper in *Bureau of Mines Information Circular 8866*, pp. 70–83, U.S. Bureau of Mines, Pittsburgh, PA.

19. Kivade, S.B., Murthy, Ch. S.N., and Vardhan, H. April, 2012. The use of Dimensional Analysis and optimization of pneumatic drilling operations and operating parameters. *J. Inst. Eng. (India), Series-D.* 93(1): 31–36.

20. Murrel, K.F.H. 1986. *Ergonomics, Man in His Working Environment*, Chapman and Hall, London.

21. Deshmukh. 1999. "Dynamics of a torsionally Flexible Clutch", M.E. (by Research) Thesis of Nagpur University under the supervision of Dr. J. P. Modak.

22. Pattiwar, J.P. 2000. "Advancement in the development of finger type torsionally flexibel clutch for a low capacity manually energized chemical unit operation device", Ph.D Thesis of Nagpur University under the Guidance of Dr. J. P. Modak.

23. Gallagher, S. and Bobick, T.G. 1988. Effects of posture on the metabolic expenditure required to lift a 50-pound box, in *Trends in Ergonomics/Human Factors*, Ed. F. Aghazadeh, pp. 927–934, Elsevier Science Publishing Co., Amsterdam.

24. Wyndham, C.H., Strydom, N.B., Benade, A.J.S., and Rensburg, A.J. 1973. Limiting rates of work for acclimatization at high wet bulb temperatures. *J. Appl. Physiol.* 35: 454.

25. Gallagher, S., Mayton, A.G., Unger, R.L., Hamrick, C.A., and Sonier, P. 1996. Computer design/evaluation tool for illuminating underground coal mining equipment. *J. Illum. Eng. Soc.* 25(1): 3–12.

26. Bedford, T. and Warner, C.G. 1955. The energy expended while working in stooping postures. *Br. J. Ind.Med.* 12: 290–295.

27. Belkhode, P.N. 2017. Mathematical modeling of liner piston maintenance activity using field data to minimize overhauling time and human energy consumption. *J. Inst. Engineers (India): Series C* Springer Publication. 99(6): 1–9.

28. Belkhode, P.N. and Bejalwar, A. 2018. Analysis of experimental setup of a small solar chimney power plant. *Elsevier Procedia Manufacturing.* 20: 481–486.

29. Belkhode, P.N. and Bejalwar, A. Sept 2019. Evaluation of the experimental data to determine the performance of a solar chimney power plant. *Material Today Proceedings.* 27(1):102–106.

30. Sivanandam, S.N., Sumathi, S., and Deepa, S.N. 2006. *Introduction to Neural Networks Using MATLAB 6.0*, Publication: Tata Mcgraw-Hill Publishing Company Limited, New Delhi.

31. Stamatios V. Kartalopoulos. 2004. *Understanding Neural Networks and Fuzzy Logic – Basic Concepts and Applications*, Publication - Prentice Hall of India Private Limited, New Delhi.

32. Rao, S.S. 1994. *Optimization Theory &Applications*, Wiley Eastern Limited, New Delhi.

Appendix

1. Computation of the Term Π_1 of Face Drilling Operation
1a. Anthropometric Dimensions of the Drilling Operator

S. No.	Age	Weight	Stature (a) mm	Shoulder Height (b) mm	Elbow Height (c) mm	Eye Height (d) mm	Finger tip Height (e) mm	Shoulder Breadth (f) mm	Hip Breadth (g) mm	Hand Breadth across Thumb (h)	Walking Length (Wl) mm	Walking Breadth (Ww) mm	$A_1 = (a*c*e*g*Wl)/(b*d*f*h*Ww)$
1	47	65	1750	1425	1080	1665	660	470	370	123	590	430	4.834772707
2	45	58	1730	1405	1085	1650	640	460	365	120	585	420	4.968331746
3	36	55	1770	1440	1090	1620	660	475	375	124	600	435	4.496127621
4	52	66	1740	1420	1060	1690	650	480	470	123	580	430	5.70156459
5	42	62	1710	1400	1020	1660	635	470	460	119	570	390	6.095939307
6	40	82	1850	1549	1030	1753	686	470	406	127	620	400	5.311571018
7	45	60	1650	1240	890	1475	735	425	355	115	610	385	5.902393016
8	53	61	1680	1260	950	1490	630	430	340	117	615	390	5.707576372
9	50	60	1700	1270	1020	1520	640	457	355	120	617	410	5.600334563
10	47	65	1750	1425	1080	1665	660	470	370	120	590	430	4.910074332
11	48	61	1640	1325	1006	1527	602	420	320	116	585	430	4.386453339
12	50	65	1750	1427	1080	1638	660	462	360	125	590	435	4.623316968
13	47	65	1750	1427	1080	1638	660	462	360	125	590	435	4.55934333
14	45	61	1640	1325	1006	1527	602	420	320	116	585	430	4.386453339
15	49	65	1750	1427	1080	1638	660	462	360	125	590	435	4.496127621
16	47	61	1640	1325	1006	1527	602	420	320	116	585	430	4.386453339
17	46	65	1750	1425	1080	1665	660	470	370	120	590	430	4.842442266
18	52	61	1680	1260	950	1490	630	430	340	117	590	390	5.341928101
19	46	60	1650	1240	890	1475	735	425	355	115	590	385	5.768039253
20	50	60	1700	1270	1020	1520	640	457	355	120	617	410	5.576006469
21	41	66	1710	1400	1020	1660	635	470	460	119	570	390	6.007937499
22	51	71	1740	1420	1060	1690	650	480	470	123	580	430	4.645237199
23	37	55	1770	1440	1090	1620	660	475	375	124	600	435	4.509558611
24	44	58	1730	1405	1085	1650	640	460	365	120	585	420	4.9691354
25	46	65	1750	1425	1080	1665	660	470	370	123	590	430	4.50943144

(*Continued*)

S. No.	Age	Weight	Stature (a) mm	Shoulder Height (b) mm	Elbow Height (c) mm	Eye Height (d) mm	Finger tip Height (e) mm	Shoulder Breadth (f) mm	Hip Breadth (g) mm	Hand Breadth across Thumb (h)	Walking Length (Wl) mm	Walking Breadth (Ww) mm	$A_1=(a*c*e*g*Wl)/(b*d*f*h*Ww)$
26	50	60	1700	1270	1020	1520	640	457	355	120	617	410	5.594779445
27	42	58	1730	1405	1085	1650	640	460	365	120	585	420	5.080519882
28	35	69	1770	1440	1090	1620	660	475	375	124	600	435	4.436831793
29	49	66	1740	1420	1060	1690	650	480	470	123	580	430	5.798708669
30	39	58	1710	1400	1020	1660	635	470	460	119	570	390	6.012170147
30	47	65	1750	1427	1080	1638	660	462	360	125	590	435	4.834772707

1b. Pi term (π_1) of Anthropometric Dimensions of the Drilling Operator

S.No.	M. No.	Age Am	Weight A2	Experience	Skill A3	Posture A4	Enthusiasm A5	Habits A6	Health A7	Anthropometric Data-A1	$\pi_1 = [(N * A_2 * A_4 * A_6 * A_1)/(A_m A_3 A_5 * A_7)]$
1	1	47	65	20	9	1	5	6	7	4.834772707	0.039187621
2	2	45	58	18	8	1	6	7	8	4.968331746	0.036227419
3	3	48	53	21	9	1	7	6	7	4.496127621	0.026762664
4	4	52	66	25	9	2	4	6	6	5.70156459	0.152285379
5	5	42	63	15	7	2	7	8	8	6.095939307	0.088862089
6	1	40	82	13	6	2	7	8	8	5.311571018	0.082202885
7	2	45	60	18	8	2	6	7	7	5.902393016	0.098373217
8	3	53	61	26	9	1	4.5	6	6	5.707576372	0.069134398
9	4	50	64	23	9	1	5	6	6	5.600334563	0.057247864
10	5	47	62	20	9	1	6	7	5	4.910074332	0.054169457
11	1	48	61	21	9	1	6	6	7	4.386453339	0.030461482
12	2	50	65	23	9	1	4	6	6	4.623316968	0.059075717

(*Continued*)

S.No.	M. No.	Age Am	Weight A2	Experience	Skill A3	Posture A4	Enthusiasm A5	Habits A6	Health A7	Anthropometric Data-A1	$\pi_1 = [(N * A_2 * A_4 * A_6 * A_1)/(A_m A_3 * A_5 * A_7)]$
13	3	47	55	20	9	1	5	7	7	4.55934333	0.043114358
14	4	45	57	18	8	2	7	7	5	4.386453339	0.087729067
15	5	48	56	21	9	2	6	6	7	4.496127621	0.062446217
16	1	47	61	20	9	2	6	7	5	4.386453339	0.096785416
17	2	46	65	19	8	2	7	7	7	4.842442266	0.071433543
18	3	52	61	25	9	1	4	6	6	5.341928101	0.071339852
19	4	46	60	19	8	1	6	7	7	5.768039253	0.049634396
20	5	50	59	23	9	1	5	6	6	5.576006469	0.056999177
21	1	41	64	14	6	1	7	8	8	6.007937499	0.04884502
22	2	48	54	21	9	1	5	6	7	4.645237199	0.03871031
23	3	45	57	18	8	1	6	7	5	4.509558611	0.052611517
24	4	44	58	17	7	2	7	7	8	4.9691354	0.068567615
25	5	48	55	21	9	2	4.5	6	7	4.50943144	0.08350799
26	1	50	60	23	9	2	4.5	6	6	5.594779445	0.127091286
27	2	42	58	15	7	2	7	8	8	5.080519882	0.074060057
28	3	48	61	21	9	1	5	6	5	4.436831793	0.051763038
29	4	49	62	22	9	1	5	6	7	5.798708669	0.049590511
30	5	39	58	12	6	1	7	8	8	6.012170147	0.044045203

2. Computation of the Term Π_2 of Face Drilling Operation

S.No.	Length of Drill Rod (Lr)	Diameter of Comp. Air Hose (Dc)	Length of Comp. Air Hose (Lc)	Diameter of Drill Rod (Dr)	[Lr * Dc * Lc]/[Dr³]
1	0.800	0.019	5.0	0.034	0.000517158
2	0.800	0.019	7.0	0.034	0.000369398
3	0.800	0.019	8.0	0.034	0.000323224
4	0.800	0.019	9.0	0.034	0.00028731
5	0.800	0.019	10.0	0.034	0.000258579
6	1.200	0.019	5.0	0.033	0.000315237
7	1.200	0.019	7.0	0.033	0.000225169
8	1.200	0.019	8.0	0.033	0.000197023
9	1.200	0.019	9.0	0.033	0.000175132
10	1.200	0.019	10.0	0.033	0.000157618
11	1.600	0.019	5.0	0.033	0.000236428

(Continued)

S.No.	Length of Drill Rod (Lr)	Diameter of Comp. Air Hose (Dc)	Length of Comp. Air Hose (Lc)	Diameter of Drill Rod (Dr)	[Lr * Dc * Lc]/[Dr³]
12	1.600	0.019	7.0	0.033	0.000168877
13	1.600	0.019	8.0	0.033	0.000147767
14	1.600	0.019	9.0	0.033	0.000131349
15	1.600	0.019	10.0	0.033	0.000118214
16	0.800	0.019	6.0	0.034	0.000430965
17	0.800	0.019	8.0	0.034	0.000323224
18	0.800	0.019	9.0	0.034	0.00028731
19	0.800	0.019	10.0	0.034	0.000258579
20	0.800	0.019	12.0	0.034	0.000215482
21	1.200	0.019	6.0	0.033	0.000262697
22	1.200	0.019	8.0	0.033	0.000197023
23	1.200	0.019	9.0	0.033	0.000175132
24	1.200	0.019	10.0	0.033	0.000157618
25	1.200	0.019	12.0	0.033	0.000131349
26	1.600	0.019	6.0	0.033	0.000197023
27	1.600	0.019	8.0	0.033	0.000147767
28	1.600	0.019	9.0	0.033	0.000131349
29	1.600	0.019	10.0	0.033	0.000118214
30	1.600	0.019	12.0	0.033	9.85115E−05

3. Computation of the Term Π_3 of Face Drilling Operation

S.No.	Wt. of Drill Rod (Wr)	Wt. of Comp. Air Hose (Wc)	Wt. of Jack Hammer (Wj)	Dia. of Drill Rod (Dr)	Shear Strength of Ore ((So) (N/m²)	Shear Strength of Mica Schist (Ss) (N/m²)	Density of Ore (Do) Kg/m³	Ambient Air Velocity (Ar)	Density of Mica Schist (Ds) Kg/m³	Comp Air Pressure (Pa)	Hardness of Drill Rod (Hr)	Rate of Water Flow thro' Hose (Qw)	$\pi_3 = [(Wr * Wc * Wj)/(Dr^2 * Pa)^3] * [\{(So * Ss * Hr)/(Pa)\}/(Pa)]*\{(Do * R^2)/(Pa)^3\} * \{(Ds * R^2)/(Pa)\}]$
1	2.9	0.9	25	0.034	556,0000	4,008,000	3500	0.8	2300	6.0	56	1.5714E−05	2.9403891
2	2.9	1.08	25	0.034	556,0000	4,008,000	3500	0.9	2300	5.9	56	1.5714E−05	3.4835941
3	2.9	1.2	25	0.034	556,0000	4,008,000	3500	1	2300	5.8	56	1.5714E−05	3.8087295
4	2.9	1.32	25	0.034	556,0000	4,008,000	3500	1.1	2300	6.1	56	1.5714E−05	4.5611273
5	2.9	1.5	25	0.034	556,0000	4,008,000	3500	1.2	2300	6.0	56	1.5714E−05	4.6687068

(Continued)

S.No.	Wt. of Drill Rod (Wr)	Wt. of Comp. Air Hose (Wc)	Wt. of Jack Hammer (Wj)	Dia. of Drill Rod (Dr)	Shear Strength of Ore ((So) (N/m²)	Shear Strength of Mica Schist (Ss) (N/m²)	Density of Ore (Do) Kg/m³	Ambient Air Velocity (Ar)	Density of Mica Schist (Ds) Kg/m³	Comp Air Pressure (Pa)	Hardness of Drill Rod (Hr)	Rate of Water Flow thro' Hose (Qw)	$\pi_3 = [(Wr * Wc * Wj)/(Dr^2 * Pa)^3] * [\{(So * Ss * Hr)/(Pa)\}*\{(Do * R^2)/(Pa)^3\}*\{(Ds * R^2)/(Pa)\}]$
6	4.1	0.9	25	0.033	556,0000	4,008,000	3500	0.8	2300	6.0	56	1.5714E-05	4.9725657
7	4.1	1.08	25	0.033	556,0000	4,008,000	3500	0.9	2300	5.9	56	1.5714E-05	6.3456155
8	4.1	1.2	25	0.033	556,0000	4,008,000	3500	1	2300	5.8	56	1.5714E-05	7.3191638
9	4.1	1.32	25	0.033	556,0000	4,008,000	3500	1.1	2300	6.1	56	1.5714E-05	7.1950756
10	4.1	1.5	25	0.033	556,0000	4,008,000	3500	1.2	2300	6.0	56	1.5714E-05	8.9031511
11	5.2	0.9	25	0.033	556,0000	4,008,000	3500	0.8	2300	6.0	56	1.5714E-05	6.3066687
12	5.2	1.08	25	0.033	556,0000	4,008,000	3500	0.9	2300	5.9	56	1.5714E-05	8.2474219
13	5.2	1.2	25	0.033	556,0000	4,008,000	3500	1	2300	5.8	56	1.5714E-05	9.7587897
14	5.2	1.32	25	0.033	556,0000	4,008,000	3500	1.1	2300	6.1	56	1.5714E-05	8.9138659
15	5.2	1.5	25	0.033	556,0000	4,008,000	3500	1.2	2300	6.0	56	1.5714E-05	10.766692
16	2.9	0.9	25	0.034	556,0000	4,008,000	3500	0.8	2300	6.0	56	1.5714E-05	2.9403891
17	2.9	1.08	25	0.034	556,0000	4,008,000	3500	0.9	2300	5.9	56	1.5714E-05	3.4835941
18	2.9	1.2	25	0.034	556,0000	4,008,000	3500	1	2300	5.8	56	1.5714E-05	3.8087295
19	2.9	1.32	25	0.034	556,0000	4,008,000	3500	1.1	2300	6.1	56	1.5714E-05	4.5611273
20	2.9	1.5	25	0.034	556,0000	4,008,000	3500	1.2	2300	6.0	56	1.5714E-05	4.6687068
21	4.1	0.9	25	0.033	556,0000	4,008,000	3500	0.8	2300	6.0	56	1.5714E-05	4.9725657
22	4.1	1.08	25	0.033	556,0000	4,008,000	3500	0.9	2300	5.9	56	1.5714E-05	6.3456155
23	4.1	1.2	25	0.033	556,0000	4,008,000	3500	1	2300	5.8	56	1.5714E-05	7.3191638
24	4.1	1.32	25	0.033	556,0000	4,008,000	3500	1.1	2300	6.1	56	1.5714E-05	7.1950756
25	4.1	1.5	25	0.033	556,0000	4,008,000	3500	1.2	2300	6.0	56	1.5714E-05	8.9031511
26	5.2	0.9	25	0.033	556,0000	4,008,000	3500	0.8	2300	6.0	56	1.5714E-05	6.3066687
27	5.2	1.08	25	0.033	556,0000	4,008,000	3500	0.9	2300	5.9	56	1.5714E-05	8.2474219
28	5.2	1.2	25	0.033	556,0000	4,008,000	3500	1	2300	5.8	56	1.5714E-05	9.7587897
29	5.2	1.32	25	0.033	556,0000	4,008,000	3500	1.1	2300	6.1	56	1.5714E-05	8.9138659
30	5.2	1.5	25	0.033	556,0000	4,008,000	3500	1.2	2300	6.0	56	1.5714E-05	10.766692

4. Computation of the Term Π_4 of Face Drilling Operation

S.No.	Dia. Drill Rod (Dr)	Speed of Drill M/c (N)	Penetration Rate (R)	Time (t) (s)	Air Velocity (Ar)	$\pi_4 = [(Dr * N * Ar)/(R)^2]$
1	0.034	60	0.00166	360	0.8	0.823956999
2	0.034	49	0.0016	375	0.9	0.845407732
3	0.034	46	0.00154	390	1	0.988970286
4	0.034	64	0.00174	345	1.1	1.049330964
5	0.034	66	0.00164	365	1.2	1.409876954
6	0.033	60	0.00166	600	0.8	0.848925393
7	0.033	56	0.00163	615	0.9	0.941499892
8	0.033	69	0.00159	630	1	1.388705863
9	0.033	62	0.00171	585	1.1	1.103439673
10	0.033	76	0.00169	590	1.2	1.528576842
11	0.033	58	0.00166	840	0.8	0.82062788
12	0.033	60	0.00164	855	0.9	0.99040943
13	0.033	55	0.00161	870	1	1.066197384
14	0.033	62	0.0017	825	1.1	1.123026907
15	0.033	66	0.00167	835	1.2	1.375714514
16	0.034	60	0.00166	360	0.8	0.823956999
17	0.034	49	0.0016	375	0.9	0.845407732
18	0.034	46	0.00154	390	1	0.988970286
19	0.034	64	0.00174	345	1.1	1.049330964
20	0.034	66	0.00164	365	1.2	1.409876954
21	0.033	60	0.00166	600	0.8	0.848925393
22	0.033	56	0.00163	615	0.9	0.941499892
23	0.033	69	0.00159	630	1	1.388705863
24	0.033	62	0.00171	585	1.1	1.103439673
25	0.033	76	0.00169	590	1.2	1.528576842
26	0.033	58	0.00166	840	0.8	0.82062788
27	0.033	60	0.00164	855	0.9	0.99040943
28	0.033	55	0.00161	870	1	1.066197384
29	0.033	62	0.0017	825	1.1	1.123026907
30	0.033	66	0.00167	835	1.2	1.375714514

2300 strokes/ Min = 38 strokes/s
38 * 9.45 Degrees = 1 Rev./ s. = 60 Rev./ min
1 M length Hole = 300 s.
Feed rate = Penetration rate = 1/300 = 0.033 m/s

5. Computation of the Term Π_5 of Face Drilling Operation

S.No.	Ambient Temp. (°C)	Relative Humidity	$\pi5 = [(\theta/100) * \emptyset]$
1	27	90	24.3
2	27	90	24.3
3	27	90	24.3
4	27	91	24.57
5	27	91	24.57
6	28	91	25.48
7	28	91	25.48
8	28	92	25.76
9	28	92	25.76
10	28	92	25.76
11	29	93	26.97
12	29	93	26.97
13	29	93	26.97
14	29	89	25.81
15	29	89	25.81
16	27	89	24.03
17	27	88	23.76
18	27	88	23.76
19	27	88	23.76
20	27	87	23.49
21	28	87	24.36
22	28	87	24.36
23	28	86	24.08
24	28	86	24.08
25	28	86	24.08
26	29	85	24.65
27	29	85	24.65
28	29	85	24.65
29	29	90	26.1
30	29	91	26.39

6. Computation of the Term Π_6 of Face Drilling Operation

S.No.	Illumination (I)	Ambient.Air Velocity (Ar)	Compressed Air Pressure (Pa)	Penetration Rate[R]	$\pi6 = I /[Pa * R]$
1	1615	0.8	6	0.00166	1.65289087
2	1615	0.9	5.9	0.0016	1.743939944
3	1615	1	5.8	0.00154	1.843125064
4	1615	1.1	6.1	0.00174	1.551045136
5	1615	1.2	6	0.00164	1.673048076
6	1615	0.8	6	0.00166	1.65289087
7	1615	0.9	5.9	0.00163	1.711842889
8	1615	1	5.8	0.00159	1.785165156
9	1615	1.1	6.1	0.00171	1.578256454
10	1615	1.2	6	0.00169	1.623549612
11	1615	0.8	6	0.00166	1.65289087
12	1615	0.9	5.9	0.00164	1.701404823
13	1615	1	5.8	0.00161	1.762989191
14	1615	1.1	6.1	0.0017	1.587540315
15	1615	1.2	6	0.00167	1.64299332
16	1615	0.8	6	0.00166	1.65289087
17	1615	0.9	5.9	0.0016	1.743939944
18	1615	1	5.8	0.00154	1.843125064
19	1615	1.1	6.1	0.00174	1.551045136
20	1615	1.2	6	0.00164	1.673048076
21	1615	0.8	6	0.00166	1.65289087
22	1615	0.9	5.9	0.00163	1.711842889
23	1615	1	5.8	0.00159	1.785165156
24	1615	1.1	6.1	0.00171	1.578256454
25	1615	1.2	6	0.00169	1.623549612
26	1615	0.8	6	0.00166	1.65289087
27	1615	0.9	5.9	0.00164	1.701404823
28	1615	1	5.8	0.00161	1.762989191
29	1615	1.1	6.1	0.0017	1.587540315
30	1615	1.2	6	0.00167	1.64299332

7. Computation of the Term Π_{D1} of Face Drilling Operation

		Time of Drilling		
S.No.	Time of Drilling (s)	Penetration Rate (R)	Dia. of Drill Rod	Z1 = [Td * R/Dr]
1	360	0.00166	0.034	17.5764706
2	375	0.0016	0.034	17.6470588
3	390	0.00154	0.034	17.6647059
4	345	0.00174	0.034	17.6558824
5	365	0.00164	0.034	17.6058824
6	600	0.00166	0.033	30.1818182
7	615	0.00163	0.033	30.3772727
8	630	0.00159	0.033	30.3545455
9	585	0.00171	0.033	30.3136364
10	590	0.00169	0.033	30.2151515
11	840	0.00166	0.033	42.2545455
12	855	0.00164	0.033	42.4909091
13	870	0.00161	0.033	42.4454545
14	825	0.0017	0.033	42.5
15	835	0.00167	0.033	42.2560606
16	360	0.00166	0.034	17.5764706
17	375	0.0016	0.034	17.6470588
18	390	0.00154	0.034	17.6647059
19	345	0.00174	0.034	17.6558824
20	365	0.00164	0.034	17.6058824
21	600	0.00166	0.033	30.1818182
22	615	0.00163	0.033	30.3772727
23	630	0.00159	0.033	30.3545455
24	585	0.00171	0.033	30.3136364
25	590	0.00169	0.033	30.2151515
26	840	0.00166	0.033	42.2545455
27	855	0.00164	0.033	42.4909091
28	870	0.00161	0.033	42.4454545
29	825	0.0017	0.033	42.5
30	835	0.00167	0.033	42.2560606

8. Computation of the Term ΠD_2 of Face Drilling Operation

	Productivity of Drilling			
S.No.	Productivity (Pd)	Dia. of Drill Rod (Dr)	Penetration Rate (R)	$Z_2 = [Pd * Dr/R]$
1	0.09027	0.034	0.00166	1.848903614
2	0.09444	0.034	0.0016	2.00685
3	0.09722	0.034	0.00154	2.146415584
4	0.10069	0.034	0.00174	1.967505747
5	0.10416	0.034	0.00164	2.159414634
6	0.09027	0.033	0.00166	1.794524096
7	0.09444	0.033	0.00163	1.91197546
8	0.09722	0.033	0.00159	2.017773585
9	0.10069	0.033	0.00171	1.943140351
10	0.10416	0.033	0.00169	2.033893491
11	0.09027	0.033	0.00166	1.794524096
12	0.09444	0.033	0.00164	1.900317073
13	0.09722	0.033	0.00161	1.992708075
14	0.10069	0.033	0.0017	1.954570588
15	0.10416	0.033	0.00167	2.058251497
16	0.09027	0.034	0.00166	1.848903614
17	0.09444	0.034	0.0016	2.00685
18	0.09722	0.034	0.00154	2.146415584
19	0.10069	0.034	0.00174	1.967505747
20	0.10416	0.034	0.00164	2.159414634
21	0.09027	0.033	0.00166	1.794524096
22	0.09444	0.033	0.00163	1.91197546
23	0.09722	0.033	0.00159	2.017773585
24	0.10069	0.033	0.00171	1.943140351
25	0.10416	0.033	0.00169	2.033893491
26	0.09027	0.033	0.00166	1.794524096
27	0.09444	0.033	0.00164	1.900317073
28	0.09722	0.033	0.00161	1.992708075
29	0.10069	0.033	0.0017	1.954570588
30	0.10416	0.033	0.00167	2.058251497

9. Computation of the Term ΠD_3 of Face Drilling Operation

	Human Energy			
S.No.	Human Energy KJ He)	Dia. of Drill Rod (Dr)	Comp. Air Pr (Pa)	$Z_3 = [He/(Dr^3 * Pa)]$
1	75.335	0.034	6	3.256415279
2	136.1663	0.034	5.9	5.985657551
3	113.8725	0.034	5.8	5.091961407
4	72.1338	0.034	6.1	3.066925439
5	128.1612	0.034	6	5.539869779
6	186.2754	0.033	6	8.806299006
7	120.66	0.033	5.9	5.800968168
8	180.4045	0.033	5.8	8.822842891
9	175.7925	0.033	6.1	8.174471432
10	210.3212	0.033	6	9.943080914
11	144.9224	0.033	6	6.851307189
12	173.4538	0.033	5.9	8.33913453
13	247.9265	0.033	5.8	12.12506649
14	245.685	0.033	6.1	11.42452046
15	298.7797	0.033	6	14.12501798
16	74.3812	0.034	6	3.215186515
17	138.1575	0.034	5.9	6.073187588
18	117.0185	0.034	5.8	5.232639012
19	69.3703	0.034	6.1	2.949429224
20	129.64	0.034	6	5.603792085
21	180.3172	0.033	6	8.52462096
22	127.6238	0.033	5.9	6.135766627
23	171.1225	0.033	5.8	8.368898407
24	172.6577	0.033	6.1	8.028701089
25	209.924	0.033	6	9.924303008
26	276.587	0.033	6	13.07584267
27	164.889	0.033	5.9	7.927364829
28	243.2134	0.033	5.8	11.89456814
29	250.6386	0.033	6.1	11.65486625
30	289.2941	0.033	6	13.67657965

10. Time for Face Drilling (Independent π Terms)

P1	P2	P3	P4	P5	P6
0.039187621	0.000517158	2.9403891	0.823957	24.3	1.652891
0.036227419	0.000369398	3.4835941	0.84540773	24.3	1.74394
0.026762664	0.000323224	3.8087295	0.98897029	24.3	1.843125
0.152285379	0.00028731	4.5611273	1.04933096	24.57	1.551045
0.088862089	0.000258579	4.6687068	1.40987695	24.57	1.673048
0.082202885	0.000315237	4.9725657	0.84892539	25.48	1.652891
0.098373217	0.000225169	6.3456155	0.94149989	25.48	1.711843
0.069134398	0.000197023	7.3191638	1.38870586	25.76	1.785165
0.057247864	0.000175132	7.1950756	1.10343967	25.76	1.578256
0.054169457	0.000157618	8.9031511	1.52857684	25.76	1.62355
0.030461482	0.000236428	6.3066687	0.82062788	26.97	1.652891
0.059075717	0.000168877	8.2474219	0.99040943	26.97	1.701405
0.043114358	0.000147767	9.7587897	1.06619738	26.97	1.762989
0.087729067	0.000131349	8.9138659	1.12302691	25.81	1.58754
0.062446217	0.000118214	10.766692	1.37571451	25.81	1.642993
0.096785416	0.000430965	2.9403891	0.823957	24.03	1.652891
0.071433543	0.000323224	3.4835941	0.84540773	23.76	1.74394
0.071339852	0.00028731	3.8087295	0.98897029	23.76	1.843125
0.049634396	0.000258579	4.5611273	1.04933096	23.76	1.551045
0.056999177	0.000215482	4.6687068	1.40987695	23.49	1.673048
0.04884502	0.000262697	4.9725657	0.84892539	24.36	1.652891
0.03871031	0.000197023	6.3456155	0.94149989	24.36	1.711843
0.052611517	0.000175132	7.3191638	1.38870586	24.08	1.785165
0.068567615	0.000157618	7.1950756	1.10343967	24.08	1.578256
0.08350799	0.000131349	8.9031511	1.52857684	24.08	1.62355
0.127091286	0.000197023	6.3066687	0.82062788	24.65	1.652891
0.074060057	0.000147767	8.2474219	0.99040943	24.65	1.701405
0.051763038	0.000131349	9.7587897	1.06619738	24.65	1.762989
0.049590511	0.000118214	8.9138659	1.12302691	26.1	1.58754
0.044045203	9.85115E-05	10.766692	1.37571451	26.39	1.642993

11. Time for Face Drilling (Dependent π Term-PD1)

Sr. No.	PD3 = He/Dr³ * Pa	PD2 = Pd * Dr/Ar	Z1 = PD1 = (Tl * Ar/D)
1	3.256415279	1.848903614	17.57647059
2	5.985657551	2.00685	17.64705882
3	5.091961407	2.146415584	17.66470588
4	3.066925439	1.967505747	17.65588235
5	5.539869779	2.159414634	17.60588235
6	8.806299006	1.794524096	30.18181818
7	5.800968168	1.91197546	30.37727273
8	8.822842891	2.017773585	30.35454545
9	8.174471432	1.943140351	30.31363636
10	9.943080914	2.033893491	30.21515152
11	6.851307189	1.794524096	42.25454545
12	8.33913453	1.900317073	42.49090909
13	12.12506649	1.992708075	42.44545455
14	11.42452046	1.954570588	42.5
15	14.12501798	2.058251497	42.25606061
16	3.215186515	1.848903614	17.57647059
17	6.073187588	2.00685	17.64705882
18	5.232639012	2.146415584	17.66470588
19	2.949429224	1.967505747	17.65588235
20	5.603792085	2.159414634	17.60588235
21	8.52462096	1.794524096	30.18181818
22	6.135766627	1.91197546	30.37727273
23	8.368898407	2.017773585	30.35454545
24	8.028701089	1.943140351	30.31363636
25	9.924303008	2.033893491	30.21515152
26	13.07584267	1.794524096	42.25454545
27	7.927364829	1.900317073	42.49090909
28	11.89456814	1.992708075	42.44545455
29	11.65486625	1.954570588	42.5
30	13.67657965	2.058251497	42.25606061

12.1 Time for Face Drilling (Independent π Terms) -1

Log Z	Log π1	Log π2	Log π3	Log π4	Log π5	Log π6
1.244932	−1.40685	−3.28638	0.468405	−0.0841	1.385606	0.218244
1.246672	−1.44096	−3.4325	0.542028	−0.07293	1.385606	0.241532
1.247106	−1.57247	−3.4905	0.58078	−0.00482	1.385606	0.265555
1.246889	−0.81734	−3.54165	0.659072	0.020912	1.390405	0.190624
1.245658	−1.05128	−3.58741	0.669197	0.149181	1.390405	0.223508
1.479745	−1.08511	−3.50136	0.696581	−0.07113	1.406199	0.218244
1.482549	−1.00712	−3.64749	0.802474	−0.02618	1.406199	0.233464
1.482224	−1.16031	−3.70548	0.864461	0.14261	1.410946	0.251678
1.481638	−1.24224	−3.75664	0.857035	0.042749	1.410946	0.198178
1.480225	−1.26625	−3.80239	0.949544	0.184287	1.410946	0.210466
1.625873	−1.51625	−3.6263	0.7998	−0.08585	1.430881	0.218244
1.628296	−1.22859	−3.77243	0.916318	−0.00419	1.430881	0.230808
1.627831	−1.36538	−3.83042	0.989396	0.027838	1.430881	0.24625
1.628389	−1.05686	−3.88157	0.950066	0.05039	1.411788	0.200725
1.625889	−1.20449	−3.92733	1.032082	0.138528	1.411788	0.215636
1.244932	−1.01419	−3.36556	0.468405	−0.0841	1.380754	0.218244
1.246672	−1.1461	−3.4905	0.542028	−0.07293	1.375846	0.241532
1.247106	−1.14667	−3.54165	0.58078	−0.00482	1.375846	0.265555
1.246889	−1.30422	−3.58741	0.659072	0.020912	1.375846	0.190624
1.245658	−1.24413	−3.66659	0.669197	0.149181	1.370883	0.223508
1.479745	−1.31118	−3.58054	0.696581	−0.07113	1.386677	0.218244
1.482549	−1.41217	−3.70548	0.802474	−0.02618	1.386677	0.233464
1.482224	−1.27892	−3.75664	0.864461	0.14261	1.381656	0.251678
1.481638	−1.16388	−3.80239	0.857035	0.042749	1.381656	0.198178
1.480225	−1.07827	−3.88157	0.949544	0.184287	1.381656	0.210466
1.625873	−0.89588	−3.70548	0.7998	−0.08585	1.391817	0.218244
1.628296	−1.13042	−3.83042	0.916318	−0.00419	1.391817	0.230808
1.627831	−1.28598	−3.88157	0.989396	0.027838	1.391817	0.24625
1.628389	−1.3046	−3.92733	0.950066	0.05039	1.416641	0.200725
1.625889	−1.3561	−4.00651	1.032082	0.138528	1.421439	0.215636

12.2 Time for Face Drilling (Independent π Terms) -2

Z = Log Z	A = Log π1	B = Log π2	C = Log π3	D = Log π4	E = Log π5	F = Log π6
1.244932	−1.40685	−3.28638	0.468405	−0.0841	1.385606	0.218244
1.246672	−1.44096	−3.4325	0.542028	−0.07293	1.385606	0.241532
1.247106	−1.57247	−3.4905	0.58078	−0.00482	1.385606	0.265555
1.246889	−0.81734	−3.54165	0.659072	0.020912	1.390405	0.190624
1.245658	−1.05128	−3.58741	0.669197	0.149181	1.390405	0.223508
1.479745	−1.08511	−3.50136	0.696581	−0.07113	1.406199	0.218244
1.482549	−1.00712	−3.64749	0.802474	−0.02618	1.406199	0.233464
1.482224	−1.16031	−3.70548	0.864461	0.14261	1.410946	0.251678
1.481638	−1.24224	−3.75664	0.857035	0.042749	1.410946	0.198178
1.480225	−1.26625	−3.80239	0.949544	0.184287	1.410946	0.210466
1.625873	−1.51625	−3.6263	0.7998	−0.08585	1.430881	0.218244
1.628296	−1.22859	−3.77243	0.916318	−0.00419	1.430881	0.230808
1.627831	−1.36538	−3.83042	0.989396	0.027838	1.430881	0.24625
1.628389	−1.05686	−3.88157	0.950066	0.05039	1.411788	0.200725
1.625889	−1.20449	−3.92733	1.032082	0.138528	1.411788	0.215636
1.244932	−1.01419	−3.36556	0.468405	−0.0841	1.380754	0.218244
1.246672	−1.1461	−3.4905	0.542028	−0.07293	1.375846	0.241532
1.247106	−1.14667	−3.54165	0.58078	−0.00482	1.375846	0.265555
1.246889	−1.30422	−3.58741	0.659072	0.020912	1.375846	0.190624
1.245658	−1.24413	−3.66659	0.669197	0.149181	1.370883	0.223508
1.479745	−1.31118	−3.58054	0.696581	−0.07113	1.386677	0.218244
1.482549	−1.41217	−3.70548	0.802474	−0.02618	1.386677	0.233464
1.482224	−1.27892	−3.75664	0.864461	0.14261	1.381656	0.251678
1.481638	−1.16388	−3.80239	0.857035	0.042749	1.381656	0.198178
1.480225	−1.07827	**−3.88157**	0.949544	0.184287	1.381656	0.210466
1.625873	−0.89588	−3.70548	0.7998	−0.08585	1.391817	0.218244
1.628296	−1.13042	−3.83042	0.916318	−0.00419	1.391817	0.230808
1.627831	−1.28598	−3.88157	0.989396	0.027838	1.391817	0.24625
1.628389	−1.3046	−3.92733	0.950066	0.05039	1.416641	0.200725
1.625889	−1.3561	−4.00651	1.032082	0.138528	1.421439	0.215636
43.548	**−36.49**	**−110.5**	**23.554**	**0.8146**	**41.91**	**6.7263**

12.3 Time for Face Drilling (Independent π Terms) -3

ZA	A	AA	AB	AC	AD	AE	AF
−1.75143	−1.40685	1.97923	4.623443	−0.65898	0.11831	−1.94934	−0.30704
−1.79641	−1.44096	2.076373	4.946111	−0.78104	0.105095	−1.99661	−0.34804
−1.96104	−1.57247	2.472664	5.488704	−0.91326	0.007574	−2.17883	−0.41758
−1.01913	−0.81734	0.668048	2.894738	−0.53869	−0.01709	−1.13644	−0.15581
−1.30954	−1.05128	1.105197	3.771382	−0.70352	−0.15683	−1.46171	−0.23497
−1.60569	−1.08511	1.17747	3.799374	−0.75587	0.077185	−1.52589	−0.23682
−1.49311	−1.00712	1.014297	3.673473	−0.80819	0.026366	−1.41622	−0.23513
−1.71983	−1.16031	1.34631	4.299493	−1.00304	−0.16547	−1.63713	−0.29202
−1.84055	−1.24224	1.543162	4.666646	−1.06464	−0.0531	−1.75273	−0.24618
−1.87433	−1.26625	1.603378	4.814763	−1.20236	−0.23335	−1.7866	−0.2665
−2.46523	−1.51625	2.299011	5.498376	−1.2127	0.130176	−2.16957	−0.33091
−2.00051	−1.22859	1.509436	4.634773	−1.12578	0.005142	−1.75797	−0.28357
−2.22261	−1.36538	1.864257	5.229974	−1.3509	−0.03801	−1.95369	−0.33622
−1.72097	−1.05686	1.116946	4.102267	−1.00408	−0.05326	−1.49206	−0.21214
−1.95837	−1.20449	1.450805	4.730447	−1.24314	−0.16686	−1.70049	−0.25973
−1.2626	−1.01419	1.028582	3.413316	−0.47505	0.085289	−1.40035	−0.22134
−1.42881	−1.1461	1.31354	4.000451	−0.62122	0.083589	−1.57685	−0.27682
−1.43002	−1.14667	1.314847	4.061095	−0.66596	0.005523	−1.57764	−0.3045
−1.62621	−1.30422	1.700983	4.678758	−0.85957	−0.02727	−1.7944	−0.24862
−1.54976	−1.24413	1.547863	4.561717	−0.83257	−0.1856	−1.70556	−0.27807
−1.94021	−1.31118	1.719192	4.694737	−0.91334	0.093265	−1.81818	−0.28616
−2.09362	−1.41217	1.994234	5.232784	−1.13323	0.03697	−1.95823	−0.32969
−1.89564	−1.27892	1.635634	4.804433	−1.10558	−0.18239	−1.76703	−0.32188
−1.72445	−1.16388	1.354619	4.425533	−0.99749	−0.04975	−1.60808	−0.23066
−1.59608	−1.07827	1.16267	4.185393	−1.02387	−0.19871	−1.4898	−0.22694
−1.45659	−0.89588	0.802609	3.319684	−0.71653	0.076915	−1.24691	−0.19552
−1.84065	−1.13042	1.27784	4.32997	−1.03582	0.004731	−1.57333	−0.26091
−2.09336	−1.28598	1.653745	4.991628	−1.27234	−0.0358	−1.78985	−0.31667
−2.1244	−1.3046	1.701985	5.123603	−1.23946	−0.06574	−1.84815	−0.26187
−2.20487	−1.3561	1.839011	5.433238	−1.39961	−0.18786	−1.92762	−0.29242
−53.01	**−36.49**	**45.274**	**134.43**	**−28.66**	**−0.961**	**−51**	**−8.215**

12.4 Time for Face Drilling (Independent π Terms) -4

ZB	B	AB	BB	BC	BD	BE	BF
−4.09131	−3.28638	4.623443	10.80027	−1.53935	0.276369	−4.55362	−0.71723
−4.27921	−3.4325	4.946111	11.78209	−1.86051	0.250346	−4.7561	−0.82906
−4.35302	−3.4905	5.488704	12.18357	−2.02721	0.016813	−4.83645	−0.92692
−4.41605	−3.54165	2.894738	12.54328	−2.3342	−0.07406	−4.92433	−0.67512
−4.46868	−3.58741	3.771382	12.86949	−2.40068	−0.53517	−4.98795	−0.80182
−5.18113	−3.50136	3.799374	12.25954	−2.43898	0.249054	−4.92361	−0.76415
−5.40758	−3.64749	3.673473	13.30419	−2.92702	0.09549	−5.1291	−0.85156
−5.49235	−3.70548	4.299493	13.7306	−3.20325	−0.52844	−5.22824	−0.93259
−5.56597	−3.75664	4.666646	14.11231	−3.21957	−0.16059	−5.30041	−0.74448
−5.6284	−3.80239	4.814763	14.45819	−3.61054	−0.70073	−5.36497	−0.80027
−5.89591	−3.6263	5.498376	13.15006	−2.90032	0.311332	−5.18881	−0.79142
−6.14263	−3.77243	4.634773	14.23123	−3.45675	0.015788	−5.3979	−0.87071
−6.23528	−3.83042	5.229974	14.67213	−3.7898	−0.10663	−5.48088	−0.94324
−6.32071	−3.88157	4.102267	15.06662	−3.68775	−0.19559	−5.47996	−0.77913
−6.38541	−3.92733	4.730447	15.42393	−4.05333	−0.54405	−5.54456	−0.84687
−4.18989	−3.36556	3.413316	11.32698	−1.57644	0.283028	−4.64701	−0.73451
−4.35151	−3.4905	4.000451	12.18357	−1.89195	0.254575	−4.80239	−0.84307
−4.41681	−3.54165	4.061095	12.54328	−2.05692	0.017059	−4.87277	−0.9405
−4.4731	−3.58741	4.678758	12.86949	−2.36436	−0.07502	−4.93572	−0.68385
−4.56731	−3.66659	4.561717	13.44387	−2.45367	−0.54699	−5.02646	−0.81951
−5.29829	−3.58054	4.694737	12.8203	−2.49414	0.254686	−4.96506	−0.78143
−5.49356	−3.70548	5.232784	13.7306	−2.97355	0.097009	−5.13831	−0.8651
−5.56817	−3.75664	4.804433	14.11231	−3.24747	−0.53573	−5.19038	−0.94546
−5.63377	−3.80239	4.425533	14.45819	−3.25879	−0.16255	−5.2536	−0.75355
−5.7456	−3.88157	4.185393	15.06662	−3.68572	−0.71532	−5.363	−0.81694
−6.02465	−3.70548	3.319684	13.7306	−2.96365	0.31813	−5.15735	−0.8087
−6.23706	−3.83042	4.32997	14.67213	−3.50989	0.016031	−5.33125	−0.88409
−6.31855	−3.88157	4.991628	15.06662	−3.84041	−0.10805	−5.40244	−0.95584
−6.39522	−3.92733	5.123603	15.42393	−3.73122	−0.1979	−5.56362	−0.78831
−6.51415	−4.00651	5.433238	16.05215	−4.13505	−0.55502	−5.69502	−0.86395
−161.1	**−110.5**	**134.43**	**408.09**	**−87.63**	**−3.286**	**−154.4**	**−24.76**

12.5 Time for Face Drilling (Independent π Terms) -5

ZC	C	AC	BC	CC	CD	CE	CF
0.583132	0.468405	−0.65898	−1.53935	0.219403	−0.03939	0.649025	0.102227
0.675731	0.542028	−0.78104	−1.86051	0.293794	−0.03953	0.751037	0.130917
0.724295	0.58078	−0.91326	−2.02721	0.337306	−0.0028	0.804733	0.154229
0.82179	0.659072	−0.53869	−2.3342	0.434376	0.013783	0.916377	0.125635
0.83359	0.669197	−0.70352	−2.40068	0.447824	0.099832	0.930454	0.149571
1.030762	0.696581	−0.75587	−2.43898	0.485224	−0.04955	0.979531	0.152025
1.189706	0.802474	−0.80819	−2.92702	0.643964	−0.02101	1.128438	0.187349
1.281325	0.864461	−1.00304	−3.20325	0.747294	0.123281	1.219708	0.217566
1.269816	0.857035	−1.06464	−3.21957	0.73451	0.036637	1.20923	0.169845
1.405538	0.949544	−1.20236	−3.61054	0.901633	0.174989	1.339755	0.199846
1.300374	0.7998	−1.2127	−2.90032	0.63968	−0.06867	1.144419	0.174552
1.492037	0.916318	−1.12578	−3.45675	0.839639	−0.00384	1.311142	0.211493
1.61057	0.989396	−1.3509	−3.7898	0.978904	0.027542	1.415708	0.243638
1.547077	0.950066	−1.00408	−3.68775	0.902626	0.047874	1.341292	0.190702
1.678051	1.032082	−1.24314	−4.05333	1.065194	0.142973	1.457081	0.222554
0.583132	0.468405	−0.47505	−1.57644	0.219403	−0.03939	0.646752	0.102227
0.675731	0.542028	−0.62122	−1.89195	0.293794	−0.03953	0.745747	0.130917
0.724295	0.58078	−0.66596	−2.05692	0.337306	−0.0028	0.799064	0.154229
0.82179	0.659072	−0.85957	−2.36436	0.434376	0.013783	0.906782	0.125635
0.83359	0.669197	−0.83257	−2.45367	0.447824	0.099832	0.91739	0.149571
1.030762	0.696581	−0.91334	−2.49414	0.485224	−0.04955	0.965932	0.152025
1.189706	0.802474	−1.13323	−2.97355	0.643964	−0.02101	1.112772	0.187349
1.281325	0.864461	−1.10558	−3.24747	0.747294	0.123281	1.194389	0.217566
1.269816	0.857035	−0.99749	−3.25879	0.73451	0.036637	1.184128	0.169845
1.405538	0.949544	−1.02387	−3.68572	0.901633	0.174989	1.311943	0.199846
1.300374	0.7998	−0.71653	−2.96365	0.63968	−0.06867	1.113175	0.174552
1.492037	0.916318	−1.03582	−3.50989	0.839639	−0.00384	1.275347	0.211493
1.61057	0.989396	−1.27234	−3.84041	0.978904	0.027542	1.377058	0.243638
1.547077	0.950066	−1.23946	−3.73122	0.902626	0.047874	1.345902	0.190702
1.678051	1.032082	−1.39961	−4.13505	1.065194	0.142973	1.467042	0.222554
34.888	**23.554**	**−28.66**	**−87.63**	**19.343**	**0.8843**	**32.961**	**5.2643**

12.6 Time for Face Drilling (Independent π Terms) -6

ZD	D	AD	BD	CD	DD	DE	DF
−0.10469	−0.0841	0.11831	0.276369	−0.03939	0.007072	−0.11652	−0.01835
−0.09092	−0.07293	0.105095	0.250346	−0.03953	0.005319	−0.10106	−0.01762
−0.00601	−0.00482	0.007574	0.016813	−0.0028	2.32E−05	−0.00667	−0.00128
0.026076	0.020912	−0.01709	−0.07406	0.013783	0.000437	0.029077	0.003986
0.185829	0.149181	−0.15683	−0.53517	0.099832	0.022255	0.207422	0.033343
−0.10525	−0.07113	0.077185	0.249054	−0.04955	0.00506	−0.10002	−0.01552
−0.03881	−0.02618	0.026366	0.09549	−0.02101	0.000685	−0.03681	−0.00611
0.21138	0.14261	−0.16547	−0.52844	0.123281	0.020338	0.201215	0.035892
0.063338	0.042749	−0.0531	−0.16059	0.036637	0.001827	0.060316	0.008472
0.272787	0.184287	−0.23335	−0.70073	0.174989	0.033962	0.260019	0.038786
−0.13959	−0.08585	0.130176	0.311332	−0.06867	0.007371	−0.12285	−0.01874
−0.00681	−0.00419	0.005142	0.015788	−0.00384	1.75E−05	−0.00599	−0.00097
0.045315	0.027838	−0.03801	−0.10663	0.027542	0.000775	0.039832	0.006855
0.082055	0.05039	−0.05326	−0.19559	0.047874	0.002539	0.07114	0.010115
0.225232	0.138528	−0.16686	−0.54405	0.142973	0.01919	0.195573	0.029872
−0.10469	−0.0841	0.085289	0.283028	−0.03939	0.007072	−0.11612	−0.01835
−0.09092	−0.07293	0.083589	0.254575	−0.03953	0.005319	−0.10035	−0.01762
−0.00601	−0.00482	0.005523	0.017059	−0.0028	2.32E−05	−0.00663	−0.00128
0.026076	0.020912	−0.02727	−0.07502	0.013783	0.000437	0.028772	0.003986
0.185829	0.149181	−0.1856	−0.54699	0.099832	0.022255	0.20451	0.033343
−0.10525	−0.07113	0.093265	0.254686	−0.04955	0.00506	−0.09864	−0.01552
−0.03881	−0.02618	0.03697	0.097009	−0.02101	0.000685	−0.0363	−0.00611
0.21138	0.14261	−0.18239	−0.53573	0.123281	0.020338	0.197038	0.035892
0.063338	0.042749	−0.04975	−0.16255	0.036637	0.001827	0.059064	0.008472
0.272787	0.184287	−0.19871	−0.71532	0.174989	0.033962	0.254622	0.038786
−0.13959	−0.08585	0.076915	0.31813	−0.06867	0.007371	−0.11949	−0.01874
−0.00681	−0.00419	0.004731	0.016031	−0.00384	1.75E−05	−0.00583	−0.00097
0.045315	0.027838	−0.0358	−0.10805	0.027542	0.000775	0.038745	0.006855
0.082055	0.05039	−0.06574	−0.1979	0.047874	0.002539	0.071385	0.010115
0.225232	0.138528	−0.18786	−0.55502	0.142973	0.01919	0.19691	0.029872
1.2398	**0.8146**	**−0.961**	**−3.286**	**0.8843**	**0.2537**	**1.1424**	**0.1775**

12.7 Time for Face Drilling (Independent π Terms) -7

ZE	E	AE	BE	CE	DE	EE	EF
1.724985	1.385606	−1.94934	−4.55362	0.649025	−0.11652	1.919905	0.302401
1.727397	1.385606	−1.99661	−4.7561	0.751037	−0.10106	1.919905	0.334668
1.727998	1.385606	−2.17883	−4.83645	0.804733	−0.00667	1.919905	0.367954
1.733681	1.390405	−1.13644	−4.92433	0.916377	0.029077	1.933226	0.265045
1.731969	1.390405	−1.46171	−4.98795	0.930454	0.207422	1.933226	0.310767
2.080817	1.406199	−1.52589	−4.92361	0.979531	−0.10002	1.977397	0.306895
2.084759	1.406199	−1.41622	−5.1291	1.128438	−0.03681	1.977397	0.328297
2.091337	1.410946	−1.63713	−5.22824	1.219708	0.201215	1.990768	0.355105
2.090511	1.410946	−1.75273	−5.30041	1.20923	0.060316	1.990768	0.279618
2.088517	1.410946	−1.7866	−5.36497	1.339755	0.260019	1.990768	0.296956
2.326431	1.430881	−2.16957	−5.18881	1.144419	−0.12285	2.04742	0.312281
2.329898	1.430881	−1.75797	−5.3979	1.311142	−0.00599	2.04742	0.330258
2.329233	1.430881	−1.95369	−5.48088	1.415708	0.039832	2.04742	0.352354
2.29894	1.411788	−1.49206	−5.47996	1.341292	0.07114	1.993145	0.283381
2.295411	1.411788	−1.70049	−5.54456	1.457081	0.195573	1.993145	0.304432
1.718944	1.380754	−1.40035	−4.64701	0.646752	−0.11612	1.906481	0.301341
1.71523	1.375846	−1.57685	−4.80239	0.745747	−0.10035	1.892953	0.33231
1.715827	1.375846	−1.57764	−4.87277	0.799064	−0.00663	1.892953	0.365363
1.715528	1.375846	−1.7944	−4.93572	0.906782	0.028772	1.892953	0.26227
1.707651	1.370883	−1.70556	−5.02646	0.91739	0.20451	1.87932	0.306404
2.051929	1.386677	−1.81818	−4.96506	0.965932	−0.09864	1.922874	0.302634
2.055817	1.386677	−1.95823	−5.13831	1.112772	−0.0363	1.922874	0.323739
2.047924	1.381656	−1.76703	−5.19038	1.194389	0.197038	1.908975	0.347733
2.047115	1.381656	−1.60808	−5.2536	1.184128	0.059064	1.908975	0.273813
2.045162	1.381656	−1.4898	−5.363	1.311943	0.254622	1.908975	0.290791
2.262918	1.391817	−1.24691	−5.15735	1.113175	−0.11949	1.937154	0.303756
2.26629	1.391817	−1.57333	−5.33125	1.275347	−0.00583	1.937154	0.321242
2.265643	1.391817	−1.78985	−5.40244	1.377058	0.038745	1.937154	0.342734
2.306842	1.416641	−1.84815	−5.56362	1.345902	0.071385	2.00687	0.284355
2.311103	1.421439	−1.92762	−5.69502	1.467042	0.19691	2.02049	0.306513
60.896	**41.91**	**−51**	**−154.4**	**32.961**	**1.1424**	**58.558**	**9.3954**

12.8 Time for Face Drilling (Independent π Terms) -8

ZF	F	AF	BF	CF	DF	EF	FF
0.271699	0.218244	−0.30704	−0.71723	0.102227	−0.01835	0.302401	0.047631
0.301111	0.241532	−0.34804	−0.82906	0.130917	−0.01762	0.334668	0.058337
0.331175	0.265555	−0.41758	−0.92692	0.154229	−0.00128	0.367954	0.070519
0.237688	0.190624	−0.15581	−0.67512	0.125635	0.003986	0.265045	0.036338
0.278415	0.223508	−0.23497	−0.80182	0.149571	0.033343	0.310767	0.049956
0.322946	0.218244	−0.23682	−0.76415	0.152025	−0.01552	0.306895	0.047631
0.346122	0.233464	−0.23513	−0.85156	0.187349	−0.00611	0.328297	0.054505
0.373044	0.251678	−0.29202	−0.93259	0.217566	0.035892	0.355105	0.063342
0.293627	0.198178	−0.24618	−0.74448	0.169845	0.008472	0.279618	0.039274
0.311536	0.210466	−0.2665	−0.80027	0.199846	0.038786	0.296956	0.044296
0.354837	0.218244	−0.33091	−0.79142	0.174552	−0.01874	0.312281	0.047631
0.375823	0.230808	−0.28357	−0.87071	0.211493	−0.00097	0.330258	0.053272
0.400853	0.24625	−0.33622	−0.94324	0.243638	0.006855	0.352354	0.060639
0.326858	0.200725	−0.21214	−0.77913	0.190702	0.010115	0.283381	0.04029
0.3506	0.215636	−0.25973	−0.84687	0.222554	0.029872	0.304432	0.046499
0.271699	0.218244	−0.22134	−0.73451	0.102227	−0.01835	0.301341	0.047631
0.301111	0.241532	−0.27682	−0.84307	0.130917	−0.01762	0.33231	0.058337
0.331175	0.265555	−0.3045	−0.9405	0.154229	−0.00128	0.365363	0.070519
0.237688	0.190624	−0.24862	−0.68385	0.125635	0.003986	0.26227	0.036338
0.278415	0.223508	−0.27807	−0.81951	0.149571	0.033343	0.306404	0.049956
0.322946	0.218244	−0.28616	−0.78143	0.152025	−0.01552	0.302634	0.047631
0.346122	0.233464	−0.32969	−0.8651	0.187349	−0.00611	0.323739	0.054505
0.373044	0.251678	−0.32188	−0.94546	0.217566	0.035892	0.347733	0.063342
0.293627	0.198178	−0.23066	−0.75355	0.169845	0.008472	0.273813	0.039274
0.311536	0.210466	−0.22694	−0.81694	0.199846	0.038786	0.290791	0.044296
0.354837	0.218244	−0.19552	−0.8087	0.174552	−0.01874	0.303756	0.047631
0.375823	0.230808	−0.26091	−0.88409	0.211493	−0.00097	0.321242	0.053272
0.400853	0.24625	−0.31667	−0.95584	0.243638	0.006855	0.342734	0.060639
0.326858	0.200725	−0.26187	−0.78831	0.190702	0.010115	0.284355	0.04029
0.3506	0.215636	−0.29242	−0.86395	0.222554	0.029872	0.306513	0.046499
9.7527	**6.7263**	**−8.215**	**−24.76**	**5.2643**	**0.1775**	**9.3954**	**1.5203**

12.9 Error Calculation for Time in Face Drilling [(Experimental-Model
Calculated)/Experimental] * 100

Calculated T	Expt T	Error = ET–CT	% Error
16.2371	17.5765	1.33936	7.620202477
19.0256	17.6471	1.37854	7.811715266
18.4039	17.6647	0.73915	4.184333555
21.0472	17.6559	3.39135	19.2080654
16.9429	17.6059	0.66298	3.76567167
28.4050	30.1818	1.7768	5.88698153
33.7357	30.3773	3.35841	11.05566244
28.5315	30.3545	1.82308	6.005954517
32.6826	30.3136	2.36893	7.814723982
31.2331	30.2152	1.01796	3.369052989
37.9251	42.2545	4.3294	10.24598782
43.4007	42.4909	0.90976	2.141079127
48.6529	42.4455	6.20742	14.62445286
40.4358	42.5000	2.06421	4.856966173
41.5840	42.2561	0.67211	1.590564467
16.3049	17.5765	1.27156	7.23446305
18.9054	17.6471	1.25835	7.130656802
18.4034	17.6647	0.73865	4.181518363
20.0667	17.6559	2.41078	13.65426759
16.2004	17.6059	1.40547	7.982936501
27.1202	30.1818	3.06165	10.14403036
32.0032	30.3773	1.62593	5.352464426
27.0104	30.3545	3.34415	11.01696983
31.2243	30.3136	0.91067	3.004151103
29.9133	30.2152	0.30182	0.998892728
36.4652	42.2545	5.78939	13.70122779
40.8348	42.4909	1.65615	3.897647107
45.7607	42.4455	3.31529	7.810717444
40.1657	42.5000	2.33433	5.492533819
41.6955	42.2561	0.56055	1.326556445

13.1 Productivity for Face Drilling (Independent π Terms)-1

P1=	P2=	P3=	P4=	P5=	P6=
0.039187621	0.000517158	2.9403891	0.823956999	24.3	1.65289087
0.036227419	0.000369398	3.4835941	0.845407732	24.3	1.743939944
0.026762664	0.000323224	3.8087295	0.988970286	24.3	1.843125064
0.152285379	0.00028731	4.5611273	1.049330964	24.57	1.551045136
0.088862089	0.000258579	4.6687068	1.409876954	24.57	1.673048076
0.082202885	0.000315237	4.9725657	0.848925393	25.48	1.65289087
0.098373217	0.000225169	6.3456155	0.941499892	25.48	1.711842889
0.069134398	0.000197023	7.3191638	1.388705863	25.76	1.785165156
0.057247864	0.000175132	7.1950756	1.103439673	25.76	1.578256454
0.054169457	0.000157618	8.9031511	1.528576842	25.76	1.623549612
0.030461482	0.000236428	6.3066687	0.82062788	26.97	1.65289087
0.059075717	0.000168877	8.2474219	0.99040943	26.97	1.701404823
0.043114358	0.000147767	9.7587897	1.066197384	26.97	1.762989191
0.087729067	0.000131349	8.9138659	1.123026907	25.81	1.587540315
0.062446217	0.000118214	10.766692	1.375714514	25.81	1.64299332
0.096785416	0.000430965	2.9403891	0.823956999	24.03	1.65289087
0.071433543	0.000323224	3.4835941	0.845407732	23.76	1.743939944
0.071339852	0.00028731	3.8087295	0.988970286	23.76	1.843125064
0.049634396	0.000258579	4.5611273	1.049330964	23.76	1.551045136
0.056999177	0.000215482	4.6687068	1.409876954	23.49	1.673048076
0.04884502	0.000262697	4.9725657	0.848925393	24.36	1.65289087
0.03871031	0.000197023	6.3456155	0.941499892	24.36	1.711842889
0.052611517	0.000175132	7.3191638	1.388705863	24.08	1.785165156
0.068567615	0.000157618	7.1950756	1.103439673	24.08	1.578256454
0.08350799	0.000131349	8.9031511	1.528576842	24.08	1.623549612
0.127091286	0.000197023	6.3066687	0.82062788	24.65	1.65289087
0.074060057	0.000147767	8.2474219	0.99040943	24.65	1.701404823
0.051763038	0.000131349	9.7587897	1.066197384	24.65	1.762989191
0.049590511	0.000118214	8.9138659	1.123026907	26.1	1.587540315
0.044045203	9.85115E−05	10.766692	1.375714514	26.39	1.64299332

13.2 (a) Productivity for Face Drilling (Independent π Terms)-2

Log Z	Log π1	Log π2	Log π3	Log π4	Log π5	Log π6
0.266914	−1.40685	−3.28638	0.468405	−0.0841	1.385606	0.218244
0.302515	−1.44096	−3.4325	0.542028	−0.07293	1.385606	0.241532
0.331714	−1.57247	−3.4905	0.58078	−0.00482	1.385606	0.265555
0.293916	−0.81734	−3.54165	0.659072	0.020912	1.390405	0.190624
0.334336	−1.05128	−3.58741	0.669197	0.149181	1.390405	0.223508
0.253949	−1.08511	−3.50136	0.696581	−0.07113	1.406199	0.218244
0.281482	−1.00712	−3.64749	0.802474	−0.02618	1.406199	0.233464
0.304872	−1.16031	−3.70548	0.864461	0.14261	1.410946	0.251678
0.288504	−1.24224	−3.75664	0.857035	0.042749	1.410946	0.198178
0.308328	−1.26625	−3.80239	0.949544	0.184287	1.410946	0.210466
0.253949	−1.51625	−3.6263	0.7998	−0.08585	1.430881	0.218244
0.278826	−1.22859	−3.77243	0.916318	−0.00419	1.430881	0.230808
0.299444	−1.36538	−3.83042	0.989396	0.027838	1.430881	0.24625
0.291051	−1.05686	−3.88157	0.950066	0.05039	1.411788	0.200725
0.313498	−1.20449	−3.92733	1.032082	0.138528	1.411788	0.215636
0.266914	−1.01419	−3.36556	0.468405	−0.0841	1.380754	0.218244
0.302515	−1.1461	−3.4905	0.542028	−0.07293	1.375846	0.241532
0.331714	−1.14667	−3.54165	0.58078	−0.00482	1.375846	0.265555
0.293916	−1.30422	−3.58741	0.659072	0.020912	1.375846	0.190624
0.334336	−1.24413	−3.66659	0.669197	0.149181	1.370883	0.223508
0.253949	−1.31118	−3.58054	0.696581	−0.07113	1.386677	0.218244
0.281482	−1.41217	−3.70548	0.802474	−0.02618	1.386677	0.233464
0.304872	−1.27892	−3.75664	0.864461	0.14261	1.381656	0.251678
0.288504	−1.16388	−3.80239	0.857035	0.042749	1.381656	0.198178
0.308328	−1.07827	−3.88157	0.949544	0.184287	1.381656	0.210466
0.253949	−0.89588	−3.70548	0.7998	−0.08585	1.391817	0.218244
0.278826	−1.13042	−3.83042	0.916318	−0.00419	1.391817	0.230808
0.299444	−1.28598	−3.88157	0.989396	0.027838	1.391817	0.24625
0.291051	−1.3046	−3.92733	0.950066	0.05039	1.416641	0.200725
0.313498	−1.3561	−4.00651	1.032082	0.138528	1.421439	0.215636

13.2 (b) Productivity for Face Drilling (Independent π Terms)-2

$Z = \text{Log } Z$	$A = \text{Log } \pi 1$	$B = \text{Log } \pi 2$	$C = \text{Log } \pi 3$	$D = \text{Log } \pi 4$	$E = \text{Log } \pi 5$	$F = \text{Log } \pi 6$
0.266914	−1.40685	−3.28638	0.468405	−0.0841	1.385606	0.218244
0.302515	−1.44096	−3.4325	0.542028	−0.07293	1.385606	0.241532
0.331714	−1.57247	−3.4905	0.58078	−0.00482	1.385606	0.265555
0.293916	−0.81734	−3.54165	0.659072	0.020912	1.390405	0.190624
0.334336	−1.05128	−3.58741	0.669197	0.149181	1.390405	0.223508
0.253949	−1.08511	−3.50136	0.696581	−0.07113	1.406199	0.218244
0.281482	−1.00712	−3.64749	0.802474	−0.02618	1.406199	0.233464
0.304872	−1.16031	−3.70548	0.864461	0.14261	1.410946	0.251678
0.288504	−1.24224	−3.75664	0.857035	0.042749	1.410946	0.198178
0.308328	−1.26625	−3.80239	0.949544	0.184287	1.410946	0.210466
0.253949	−1.51625	−3.6263	0.7998	−0.08585	1.430881	0.218244
0.278826	−1.22859	−3.77243	0.916318	−0.00419	1.430881	0.230808
0.299444	−1.36538	−3.83042	0.989396	0.027838	1.430881	0.24625
0.291051	−1.05686	−3.88157	0.950066	0.05039	1.411788	0.200725
0.313498	−1.20449	−3.92733	1.032082	0.138528	1.411788	0.215636
0.266914	−1.01419	−3.36556	0.468405	−0.0841	1.380754	0.218244
0.302515	−1.1461	−3.4905	0.542028	−0.07293	1.375846	0.241532
0.331714	−1.14667	−3.54165	0.58078	−0.00482	1.375846	0.265555
0.293916	−1.30422	−3.58741	0.659072	0.020912	1.375846	0.190624
0.334336	−1.24413	−3.66659	0.669197	0.149181	1.370883	0.223508
0.253949	−1.31118	−3.58054	0.696581	−0.07113	1.386677	0.218244
0.281482	−1.41217	−3.70548	0.802474	−0.02618	1.386677	0.233464
0.304872	−1.27892	−3.75664	0.864461	0.14261	1.381656	0.251678
0.288504	−1.16388	−3.80239	0.857035	0.042749	1.381656	0.198178
0.308328	−1.07827	**−3.88157**	0.949544	0.184287	1.381656	0.210466
0.253949	−0.89588	−3.70548	0.7998	−0.08585	1.391817	0.218244
0.278826	−1.13042	−3.83042	0.916318	−0.00419	1.391817	0.230808
0.299444	−1.28598	−3.88157	0.989396	0.027838	1.391817	0.24625
0.291051	−1.3046	−3.92733	0.950066	0.05039	1.416641	0.200725
0.313498	−1.3561	−4.00651	1.032082	0.138528	1.421439	0.215636
8.8066	**−36.49**	**−110.5**	**23.554**	**0.8146**	**41.91**	**6.7263**

13.3 Productivity for Face Drilling (Independent π Terms)-3

ZA	A	AA	AB	AC	AD	AE	AF
−0.37551	−1.40685	1.97923	4.623443	−0.65898	0.11831	−1.94934	−0.30704
−0.43591	−1.44096	2.076373	4.946111	−0.78104	0.105095	−1.99661	−0.34804
−0.52161	−1.57247	2.472664	5.488704	−0.91326	0.007574	−2.17883	−0.41758
−0.24023	−0.81734	0.668048	2.894738	−0.53869	−0.01709	−1.13644	−0.15581
−0.35148	−1.05128	1.105197	3.771382	−0.70352	−0.15683	−1.46171	−0.23497
−0.27556	−1.08511	1.17747	3.799374	−0.75587	0.077185	−1.52589	−0.23682
−0.28349	−1.00712	1.014297	3.673473	−0.80819	0.026366	−1.41622	−0.23513
−0.35375	−1.16031	1.34631	4.299493	−1.00304	−0.16547	−1.63713	−0.29202
−0.35839	−1.24224	1.543162	4.666646	−1.06464	−0.0531	−1.75273	−0.24618
−0.39042	−1.26625	1.603378	4.814763	−1.20236	−0.23335	−1.7866	−0.2665
−0.38505	−1.51625	2.299011	5.498376	−1.2127	0.130176	−2.16957	−0.33091
−0.34256	−1.22859	1.509436	4.634773	−1.12578	0.005142	−1.75797	−0.28357
−0.40885	−1.36538	1.864257	5.229974	−1.3509	−0.03801	−1.95369	−0.33622
−0.3076	−1.05686	1.116946	4.102267	−1.00408	−0.05326	−1.49206	−0.21214
−0.37761	−1.20449	1.450805	4.730447	−1.24314	−0.16686	−1.70049	−0.25973
−0.2707	−1.01419	1.028582	3.413316	−0.47505	0.085289	−1.40035	−0.22134
−0.34671	−1.1461	1.31354	4.000451	−0.62122	0.083589	−1.57685	−0.27682
−0.38037	−1.14667	1.314847	4.061095	−0.66596	0.005523	−1.57764	−0.3045
−0.38333	−1.30422	1.700983	4.678758	−0.85957	−0.02727	−1.7944	−0.24862
−0.41596	−1.24413	1.547863	4.561717	−0.83257	−0.1856	−1.70556	−0.27807
−0.33297	−1.31118	1.719192	4.694737	−0.91334	0.093265	−1.81818	−0.28616
−0.3975	−1.41217	1.994234	5.232784	−1.13323	0.03697	−1.95823	−0.32969
−0.38991	−1.27892	1.635634	4.804433	−1.10558	−0.18239	−1.76703	−0.32188
−0.33578	−1.16388	1.354619	4.425533	−0.99749	−0.04975	−1.60808	−0.23066
−0.33246	−1.07827	1.16267	4.185393	−1.02387	−0.19871	−1.4898	−0.22694
−0.22751	−0.89588	0.802609	3.319684	−0.71653	0.076915	−1.24691	−0.19552
−0.31519	−1.13042	1.27784	4.32997	−1.03582	0.004731	−1.57333	−0.26091
−0.38508	−1.28598	1.653745	4.991628	−1.27234	−0.0358	−1.78985	−0.31667
−0.37971	−1.3046	1.701985	5.123603	−1.23946	−0.06574	−1.84815	−0.26187
−0.42514	−1.3561	1.839011	5.433238	−1.39961	−0.18786	−1.92762	−0.29242
−10.73	**−36.49**	**45.274**	**134.43**	**−28.66**	**−0.961**	**−51**	**−8.215**

13.4 Productivity for Face Drilling (Independent π Terms)-4

ZB	B	AB	BB	BC	BD	BE	BF
−0.87718	−3.28638	4.623443	10.80027	−1.53935	0.276369	−4.55362	−0.71723
−1.03838	−3.4325	4.946111	11.78209	−1.86051	0.250346	−4.7561	−0.82906
−1.15785	−3.4905	5.488704	12.18357	−2.02721	0.016813	−4.83645	−0.92692
−1.04095	−3.54165	2.894738	12.54328	−2.3342	−0.07406	−4.92433	−0.67512
−1.1994	−3.58741	3.771382	12.86949	−2.40068	−0.53517	−4.98795	−0.80182
−0.88917	−3.50136	3.799374	12.25954	−2.43898	0.249054	−4.92361	−0.76415
−1.0267	−3.64749	3.673473	13.30419	−2.92702	0.09549	−5.1291	−0.85156
−1.1297	−3.70548	4.299493	13.7306	−3.20325	−0.52844	−5.22824	−0.93259
−1.08381	−3.75664	4.666646	14.11231	−3.21957	−0.16059	−5.30041	−0.74448
−1.17239	−3.80239	4.814763	14.45819	−3.61054	−0.70073	−5.36497	−0.80027
−0.9209	−3.6263	5.498376	13.15006	−2.90032	0.311332	−5.18881	−0.79142
−1.05185	−3.77243	4.634773	14.23123	−3.45675	0.015788	−5.3979	−0.87071
−1.147	−3.83042	5.229974	14.67213	−3.7898	−0.10663	−5.48088	−0.94324
−1.12974	−3.88157	4.102267	15.06662	−3.68775	−0.19559	−5.47996	−0.77913
−1.23121	−3.92733	4.730447	15.42393	−4.05333	−0.54405	−5.54456	−0.84687
−0.89832	−3.36556	3.413316	11.32698	−1.57644	0.283028	−4.64701	−0.73451
−1.05593	−3.4905	4.000451	12.18357	−1.89195	0.254575	−4.80239	−0.84307
−1.17481	−3.54165	4.061095	12.54328	−2.05692	0.017059	−4.87277	−0.9405
−1.0544	−3.58741	4.678758	12.86949	−2.36436	−0.07502	−4.93572	−0.68385
−1.22587	−3.66659	4.561717	13.44387	−2.45367	−0.54699	−5.02646	−0.81951
−0.90928	−3.58054	4.694737	12.8203	−2.49414	0.254686	−4.96506	−0.78143
−1.04303	−3.70548	5.232784	13.7306	−2.97355	0.097009	−5.13831	−0.8651
−1.14529	−3.75664	4.804433	14.11231	−3.24747	−0.53573	−5.19038	−0.94546
−1.09701	−3.80239	4.425533	14.45819	−3.25879	−0.16255	−5.2536	−0.75355
−1.1968	−3.88157	4.185393	15.06662	−3.68572	−0.71532	−5.363	−0.81694
−0.941	−3.70548	3.319684	13.7306	−2.96365	0.31813	−5.15735	−0.8087
−1.06802	−3.83042	4.32997	14.67213	−3.50989	0.016031	−5.33125	−0.88409
−1.16231	−3.88157	4.991628	15.06662	−3.84041	−0.10805	−5.40244	−0.95584
−1.14306	−3.92733	5.123603	15.42393	−3.73122	−0.1979	−5.56362	−0.78831
−1.25604	−4.00651	5.433238	16.05215	−4.13505	−0.55502	−5.69502	−0.86395
−32.47	**−110.5**	**134.43**	**408.09**	**−87.63**	**−3.286**	**−154.4**	**−24.76**

13.5 Productivity for Face Drilling (Independent π Terms)-5

ZC	C	AC	BC	CC	CD	CE	CF
0.125024	0.468405	−0.65898	−1.53935	0.219403	−0.03939	0.649025	0.102227
0.163971	0.542028	−0.78104	−1.86051	0.293794	−0.03953	0.751037	0.130917
0.192653	0.58078	−0.91326	−2.02721	0.337306	−0.0028	0.804733	0.154229
0.193712	0.659072	−0.53869	−2.3342	0.434376	0.013783	0.916377	0.125635
0.223737	0.669197	−0.70352	−2.40068	0.447824	0.099832	0.930454	0.149571
0.176896	0.696581	−0.75587	−2.43898	0.485224	−0.04955	0.979531	0.152025
0.225882	0.802474	−0.80819	−2.92702	0.643964	−0.02101	1.128438	0.187349
0.26355	0.864461	−1.00304	−3.20325	0.747294	0.123281	1.219708	0.217566
0.247258	0.857035	−1.06464	−3.21957	0.73451	0.036637	1.20923	0.169845
0.292771	0.949544	−1.20236	−3.61054	0.901633	0.174989	1.339755	0.199846
0.203109	0.7998	−1.2127	−2.90032	0.63968	−0.06867	1.144419	0.174552
0.255493	0.916318	−1.12578	−3.45675	0.839639	−0.00384	1.311142	0.211493
0.296268	0.989396	−1.3509	−3.7898	0.978904	0.027542	1.415708	0.243638
0.276518	0.950066	−1.00408	−3.68775	0.902626	0.047874	1.341292	0.190702
0.323556	1.032082	−1.24314	−4.05333	1.065194	0.142973	1.457081	0.222554
0.125024	0.468405	−0.47505	−1.57644	0.219403	−0.03939	0.646752	0.102227
0.163971	0.542028	−0.62122	−1.89195	0.293794	−0.03953	0.745747	0.130917
0.192653	0.58078	−0.66596	−2.05692	0.337306	−0.0028	0.799064	0.154229
0.193712	0.659072	−0.85957	−2.36436	0.434376	0.013783	0.906782	0.125635
0.223737	0.669197	−0.83257	−2.45367	0.447824	0.099832	0.91739	0.149571
0.176896	0.696581	−0.91334	−2.49414	0.485224	−0.04955	0.965932	0.152025
0.225882	0.802474	−1.13323	−2.97355	0.643964	−0.02101	1.112772	0.187349
0.26355	0.864461	−1.10558	−3.24747	0.747294	0.123281	1.194389	0.217566
0.247258	0.857035	−0.99749	−3.25879	0.73451	0.036637	1.184128	0.169845
0.292771	0.949544	−1.02387	−3.68572	0.901633	0.174989	1.311943	0.199846
0.203109	0.7998	−0.71653	−2.96365	0.63968	−0.06867	1.113175	0.174552
0.255493	0.916318	−1.03582	−3.50989	0.839639	−0.00384	1.275347	0.211493
0.296268	0.989396	−1.27234	−3.84041	0.978904	0.027542	1.377058	0.243638
0.276518	0.950066	−1.23946	−3.73122	0.902626	0.047874	1.345902	0.190702
0.323556	1.032082	−1.39961	−4.13505	1.065194	0.142973	1.467042	0.222554
6.9208	**23.554**	**−28.66**	**−87.63**	**19.343**	**0.8843**	**32.961**	**5.2643**

13.6 Productivity for Face Drilling (Independent π Terms)-6

ZD	D	AD	BD	CD	DD	DE	DF
−0.02245	−0.0841	0.11831	0.276369	−0.03939	0.007072	−0.11652	−0.01835
−0.02206	−0.07293	0.105095	0.250346	−0.03953	0.005319	−0.10106	−0.01762
−0.0016	−0.00482	0.007574	0.016813	−0.0028	2.32E−05	−0.00667	−0.00128
0.006147	0.020912	−0.01709	−0.07406	0.013783	0.000437	0.029077	0.003986
0.049877	0.149181	−0.15683	−0.53517	0.099832	0.022255	0.207422	0.033343
−0.01806	−0.07113	0.077185	0.249054	−0.04955	0.00506	−0.10002	−0.01552
−0.00737	−0.02618	0.026366	0.09549	−0.02101	0.000685	−0.03681	−0.00611
0.043478	0.14261	−0.16547	−0.52844	0.123281	0.020338	0.201215	0.035892
0.012333	0.042749	−0.0531	−0.16059	0.036637	0.001827	0.060316	0.008472
0.056821	0.184287	−0.23335	−0.70073	0.174989	0.033962	0.260019	0.038786
−0.0218	−0.08585	0.130176	0.311332	−0.06867	0.007371	−0.12285	−0.01874
−0.00117	−0.00419	0.005142	0.015788	−0.00384	1.75E−05	−0.00599	−0.00097
0.008336	0.027838	−0.03801	−0.10663	0.027542	0.000775	0.039832	0.006855
0.014666	0.05039	−0.05326	−0.19559	0.047874	0.002539	0.07114	0.010115
0.043428	0.138528	−0.16686	−0.54405	0.142973	0.01919	0.195573	0.029872
−0.02245	−0.0841	0.085289	0.283028	−0.03939	0.007072	−0.11612	−0.01835
−0.02206	−0.07293	0.083589	0.254575	−0.03953	0.005319	−0.10035	−0.01762
−0.0016	−0.00482	0.005523	0.017059	−0.0028	2.32E−05	−0.00663	−0.00128
0.006147	0.020912	−0.02727	−0.07502	0.013783	0.000437	0.028772	0.003986
0.049877	0.149181	−0.1856	−0.54699	0.099832	0.022255	0.20451	0.033343
−0.01806	−0.07113	0.093265	0.254686	−0.04955	0.00506	−0.09864	−0.01552
−0.00737	−0.02618	0.03697	0.097009	−0.02101	0.000685	−0.0363	−0.00611
0.043478	0.14261	−0.18239	−0.53573	0.123281	0.020338	0.197038	0.035892
0.012333	0.042749	−0.04975	−0.16255	0.036637	0.001827	0.059064	0.008472
0.056821	0.184287	−0.19871	−0.71532	0.174989	0.033962	0.254622	0.038786
−0.0218	−0.08585	0.076915	0.31813	−0.06867	0.007371	−0.11949	−0.01874
−0.00117	−0.00419	0.004731	0.016031	−0.00384	1.75E−05	−0.00583	−0.00097
0.008336	0.027838	−0.0358	−0.10805	0.027542	0.000775	0.038745	0.006855
0.014666	0.05039	−0.06574	−0.1979	0.047874	0.002539	0.071385	0.010115
0.043428	0.138528	−0.18786	−0.55502	0.142973	0.01919	0.19691	0.029872
0.2812	**0.8146**	**−0.961**	**−3.286**	**0.8843**	**0.2537**	**1.1424**	**0.1775**

13.7 Productivity for Face Drilling (Independent π Terms)-7

ZE	E	AE	BE	CE	DE	EE	EF
0.369838	1.385606	−1.94934	−4.55362	0.649025	−0.11652	1.919905	0.302401
0.419167	1.385606	−1.99661	−4.7561	0.751037	−0.10106	1.919905	0.334668
0.459625	1.385606	−2.17883	−4.83645	0.804733	−0.00667	1.919905	0.367954
0.408662	1.390405	−1.13644	−4.92433	0.916377	0.029077	1.933226	0.265045
0.464863	1.390405	−1.46171	−4.98795	0.930454	0.207422	1.933226	0.310767
0.357103	1.406199	−1.52589	−4.92361	0.979531	−0.10002	1.977397	0.306895
0.39582	1.406199	−1.41622	−5.1291	1.128438	−0.03681	1.977397	0.328297
0.430158	1.410946	−1.63713	−5.22824	1.219708	0.201215	1.990768	0.355105
0.407064	1.410946	−1.75273	−5.30041	1.20923	0.060316	1.990768	0.279618
0.435034	1.410946	−1.7866	−5.36497	1.339755	0.260019	1.990768	0.296956
0.363371	1.430881	−2.16957	−5.18881	1.144419	−0.12285	2.04742	0.312281
0.398967	1.430881	−1.75797	−5.3979	1.311142	−0.00599	2.04742	0.330258
0.428468	1.430881	−1.95369	−5.48088	1.415708	0.039832	2.04742	0.352354
0.410903	1.411788	−1.49206	−5.47996	1.341292	0.07114	1.993145	0.283381
0.442593	1.411788	−1.70049	−5.54456	1.457081	0.195573	1.993145	0.304432
0.368543	1.380754	−1.40035	−4.64701	0.646752	−0.11612	1.906481	0.301341
0.416214	1.375846	−1.57685	−4.80239	0.745747	−0.10035	1.892953	0.33231
0.456387	1.375846	−1.57764	−4.87277	0.799064	−0.00663	1.892953	0.365363
0.404383	1.375846	−1.7944	−4.93572	0.906782	0.028772	1.892953	0.26227
0.458336	1.370883	−1.70556	−5.02646	0.91739	0.20451	1.87932	0.306404
0.352146	1.386677	−1.81818	−4.96506	0.965932	−0.09864	1.922874	0.302634
0.390325	1.386677	−1.95823	−5.13831	1.112772	−0.0363	1.922874	0.323739
0.421229	1.381656	−1.76703	−5.19038	1.194389	0.197038	1.908975	0.347733
0.398614	1.381656	−1.60808	−5.2536	1.184128	0.059064	1.908975	0.273813
0.426004	1.381656	−1.4898	−5.363	1.311943	0.254622	1.908975	0.290791
0.353451	1.391817	−1.24691	−5.15735	1.113175	−0.11949	1.937154	0.303756
0.388075	1.391817	−1.57333	−5.33125	1.275347	−0.00583	1.937154	0.321242
0.416771	1.391817	−1.78985	−5.40244	1.377058	0.038745	1.937154	0.342734
0.412315	1.416641	−1.84815	−5.56362	1.345902	0.071385	2.00687	0.284355
0.445619	1.421439	−1.92762	−5.69502	1.467042	0.19691	2.02049	0.306513
12.3	**41.91**	**−51**	**−154.4**	**32.961**	**1.1424**	**58.558**	**9.3954**

13.8 Productivity for Face Drilling (Independent π Terms)-8

ZF	F	AF	BF	CF	DF	EF	FF
0.058252	0.218244	−0.30704	−0.71723	0.102227	−0.01835	0.302401	0.047631
0.073067	0.241532	−0.34804	−0.82906	0.130917	−0.01762	0.334668	0.058337
0.088088	0.265555	−0.41758	−0.92692	0.154229	−0.00128	0.367954	0.070519
0.056028	0.190624	−0.15581	−0.67512	0.125635	0.003986	0.265045	0.036338
0.074727	0.223508	−0.23497	−0.80182	0.149571	0.033343	0.310767	0.049956
0.055423	0.218244	−0.23682	−0.76415	0.152025	−0.01552	0.306895	0.047631
0.065716	0.233464	−0.23513	−0.85156	0.187349	−0.00611	0.328297	0.054505
0.07673	0.251678	−0.29202	−0.93259	0.217566	0.035892	0.355105	0.063342
0.057175	0.198178	−0.24618	−0.74448	0.169845	0.008472	0.279618	0.039274
0.064892	0.210466	−0.2665	−0.80027	0.199846	0.038786	0.296956	0.044296
0.055423	0.218244	−0.33091	−0.79142	0.174552	−0.01874	0.312281	0.047631
0.064355	0.230808	−0.28357	−0.87071	0.211493	−0.00097	0.330258	0.053272
0.073738	0.24625	−0.33622	−0.94324	0.243638	0.006855	0.352354	0.060639
0.058421	0.200725	−0.21214	−0.77913	0.190702	0.010115	0.283381	0.04029
0.067601	0.215636	−0.25973	−0.84687	0.222554	0.029872	0.304432	0.046499
0.058252	0.218244	−0.22134	−0.73451	0.102227	−0.01835	0.301341	0.047631
0.073067	0.241532	−0.27682	−0.84307	0.130917	−0.01762	0.33231	0.058337
0.088088	0.265555	−0.3045	−0.9405	0.154229	−0.00128	0.365363	0.070519
0.056028	0.190624	−0.24862	−0.68385	0.125635	0.003986	0.26227	0.036338
0.074727	0.223508	−0.27807	−0.81951	0.149571	0.033343	0.306404	0.049956
0.055423	0.218244	−0.28616	−0.78143	0.152025	−0.01552	0.302634	0.047631
0.065716	0.233464	−0.32969	−0.8651	0.187349	−0.00611	0.323739	0.054505
0.07673	0.251678	−0.32188	−0.94546	0.217566	0.035892	0.347733	0.063342
0.057175	0.198178	−0.23066	−0.75355	0.169845	0.008472	0.273813	0.039274
0.064892	0.210466	−0.22694	−0.81694	0.199846	0.038786	0.290791	0.044296
0.055423	0.218244	−0.19552	−0.8087	0.174552	−0.01874	0.303756	0.047631
0.064355	0.230808	−0.26091	−0.88409	0.211493	−0.00097	0.321242	0.053272
0.073738	0.24625	−0.31667	−0.95584	0.243638	0.006855	0.342734	0.060639
0.058421	0.200725	−0.26187	−0.78831	0.190702	0.010115	0.284355	0.04029
0.067601	0.215636	−0.29242	−0.86395	0.222554	0.029872	0.306513	0.046499
1.9793	**6.7263**	**−8.215**	**−24.76**	**5.2643**	**0.1775**	**9.3954**	**1.5203**

13.9 Error Calculation for Productivity Face Drilling

Cal T	Expt T	Error = ET−CT	% Error
1.8581	1.8489	0.00918	0.496373175
1.9593	2.0069	0.04756	2.370005102
2.0913	2.1464	0.05513	2.568419669
1.8965	1.9675	0.07098	3.607803536
2.1369	2.1594	0.02248	1.041215203
1.8097	1.7945	0.01518	0.845684459
1.8840	1.9120	0.02797	1.462957691
2.0823	2.0178	0.0645	3.196722974
1.9248	1.9431	0.0183	0.941662907
2.0232	2.0339	0.01068	0.524923863
1.8246	1.7945	0.03007	1.675761851
1.9125	1.9003	0.01219	0.641415139
1.9431	1.9927	0.0496	2.489306566
1.9342	1.9546	0.02038	1.042521383
1.9994	2.0583	0.05886	2.859470863
1.9124	1.8489	0.06347	3.432606198
1.9925	2.0069	0.01436	0.715325164
2.1163	2.1464	0.03007	1.400996037
1.9308	1.9675	0.03672	1.866089018
2.1942	2.1594	0.03477	1.610335074
1.8590	1.7945	0.06446	3.592147561
1.9206	1.9120	0.00863	0.451138534
2.0943	2.0178	0.07649	3.790860257
1.9261	1.9431	0.017	0.875069841
2.0537	2.0339	0.01976	0.971736472
1.8296	1.7945	0.03511	1.956585881
1.9117	1.9003	0.01133	0.596451361
1.9365	1.9927	0.05623	2.821647116
1.9890	1.9546	0.03441	1.76070509
2.0927	2.0583	0.03445	1.673924652

14.1 Human Energy Consumed for Face Drilling (Independent Terms)-1

P1	P2	P3=	P4=	P5=	P6=
0.039187621	0.000517158	2.9403891	0.823956999	24.3	1.65289087
0.036227419	0.000369398	3.4835941	0.845407732	24.3	1.743939944
0.026762664	0.000323224	3.8087295	0.988970286	24.3	1.843125064
0.152285379	0.00028731	4.5611273	1.049330964	24.57	1.551045136
0.088862089	0.000258579	4.6687068	1.409876954	24.57	1.673048076
0.082202885	0.000315237	4.9725657	0.848925393	25.48	1.65289087
0.098373217	0.000225169	6.3456155	0.941499892	25.48	1.711842889
0.069134398	0.000197023	7.3191638	1.388705863	25.76	1.785165156
0.057247864	0.000175132	7.1950756	1.103439673	25.76	1.578256454
0.054169457	0.000157618	8.9031511	1.528576842	25.76	1.623549612
0.030461482	0.000236428	6.3066687	0.82062788	26.97	1.65289087
0.059075717	0.000168877	8.2474219	0.99040943	26.97	1.701404823
0.043114358	0.000147767	9.7587897	1.066197384	26.97	1.762989191
0.087729067	0.000131349	8.9138659	1.123026907	25.81	1.587540315
0.062446217	0.000118214	10.766692	1.375714514	25.81	1.64299332
0.096785416	0.000430965	2.9403891	0.823956999	24.03	1.65289087
0.071433543	0.000323224	3.4835941	0.845407732	23.76	1.743939944
0.071339852	0.00028731	3.8087295	0.988970286	23.76	1.843125064
0.049634396	0.000258579	4.5611273	1.049330964	23.76	1.551045136
0.056999177	0.000215482	4.6687068	1.409876954	23.49	1.673048076
0.04884502	0.000262697	4.9725657	0.848925393	24.36	1.65289087
0.03871031	0.000197023	6.3456155	0.941499892	24.36	1.711842889
0.052611517	0.000175132	7.3191638	1.388705863	24.08	1.785165156
0.068567615	0.000157618	7.1950756	1.103439673	24.08	1.578256454
0.08350799	0.000131349	8.9031511	1.528576842	24.08	1.623549612
0.127091286	0.000197023	6.3066687	0.82062788	24.65	1.65289087
0.074060057	0.000147767	8.2474219	0.99040943	24.65	1.701404823
0.051763038	0.000131349	9.7587897	1.066197384	24.65	1.762989191
0.049590511	0.000118214	8.9138659	1.123026907	26.1	1.587540315
0.044045203	9.85115E−05	10.766692	1.375714514	26.39	1.64299332

14.2 Human Energy Consumed for Face Drilling (Independent Terms)-2

Z = Log Z	A = Log π1	B = Log π2	C = Log π3	D = Log π4	E = Log π5	F = Log π6
0.51274	−1.40685	−3.28638	0.468405	−0.0841	1.385606	0.218244
0.777112	−1.44096	−3.4325	0.542028	−0.07293	1.385606	0.241532
0.706885	−1.57247	−3.4905	0.58078	−0.00482	1.385606	0.265555
0.486703	−0.81734	−3.54165	0.659072	0.020912	1.390405	0.190624
0.7435	−1.05128	−3.58741	0.669197	0.149181	1.390405	0.223508
0.944793	−1.08511	−3.50136	0.696581	−0.07113	1.406199	0.218244
0.7635	−1.00712	−3.64749	0.802474	−0.02618	1.406199	0.233464
0.945609	−1.16031	−3.70548	0.864461	0.14261	1.410946	0.251678
0.91246	−1.24224	−3.75664	0.857035	0.042749	1.410946	0.198178
0.997521	−1.26625	−3.80239	0.949544	0.184287	1.410946	0.210466
0.835773	−1.51625	−3.6263	0.7998	−0.08585	1.430881	0.218244
0.921121	−1.22859	−3.77243	0.916318	−0.00419	1.430881	0.230808
1.083684	−1.36538	−3.83042	0.989396	0.027838	1.430881	0.24625
1.057838	−1.05686	−3.88157	0.950066	0.05039	1.411788	0.200725
1.149989	−1.20449	−3.92733	1.032082	0.138528	1.411788	0.215636
0.507206	−1.01419	−3.36556	0.468405	−0.0841	1.380754	0.218244
0.783417	−1.1461	−3.4905	0.542028	−0.07293	1.375846	0.241532
0.718721	−1.14667	−3.54165	0.58078	−0.00482	1.375846	0.265555
0.469738	−1.30422	−3.58741	0.659072	0.020912	1.375846	0.190624
0.748482	−1.24413	−3.66659	0.669197	0.149181	1.370883	0.223508
0.930675	−1.31118	−3.58054	0.696581	−0.07113	1.386677	0.218244
0.787869	−1.41217	−3.70548	0.802474	−0.02618	1.386677	0.233464
0.922668	−1.27892	−3.75664	0.864461	0.14261	1.381656	0.251678
0.904645	−1.16388	−3.80239	0.857035	0.042749	1.381656	0.198178
0.9967	−1.07827	**−3.88157**	0.949544	0.184287	1.381656	0.210466
1.11647	−0.89588	−3.70548	0.7998	−0.08585	1.391817	0.218244
0.899129	−1.13042	−3.83042	0.916318	−0.00419	1.391817	0.230808
1.075349	−1.28598	−3.88157	0.989396	0.027838	1.391817	0.24625
1.066507	−1.3046	−3.92733	0.950066	0.05039	1.416641	0.200725
1.135977	−1.3561	−4.00651	1.032082	0.138528	1.421439	0.215636
25.903	**−36.49**	**−110.5**	**23.554**	**0.8146**	**41.91**	**6.7263**

14.3 Human Energy Consumed for Face Drilling (Independent Terms)-3

ZA	A	AA	AB	AC	AD	AE	AF
−0.72135	−1.40685	1.97923	4.623443	−0.65898	0.11831	−1.94934	−0.30704
−1.11979	−1.44096	2.076373	4.946111	−0.78104	0.105095	−1.99661	−0.34804
−1.11156	−1.57247	2.472664	5.488704	−0.91326	0.007574	−2.17883	−0.41758
−0.3978	−0.81734	0.668048	2.894738	−0.53869	−0.01709	−1.13644	−0.15581
−0.78163	−1.05128	1.105197	3.771382	−0.70352	−0.15683	−1.46171	−0.23497
−1.02521	−1.08511	1.17747	3.799374	−0.75587	0.077185	−1.52589	−0.23682
−0.76894	−1.00712	1.014297	3.673473	−0.80819	0.026366	−1.41622	−0.23513
−1.0972	−1.16031	1.34631	4.299493	−1.00304	−0.16547	−1.63713	−0.29202
−1.13349	−1.24224	1.543162	4.666646	−1.06464	−0.0531	−1.75273	−0.24618
−1.26311	−1.26625	1.603378	4.814763	−1.20236	−0.23335	−1.7866	−0.2665
−1.26724	−1.51625	2.299011	5.498376	−1.2127	0.130176	−2.16957	−0.33091
−1.13168	−1.22859	1.509436	4.634773	−1.12578	0.005142	−1.75797	−0.28357
−1.47964	−1.36538	1.864257	5.229974	−1.3509	−0.03801	−1.95369	−0.33622
−1.11798	−1.05686	1.116946	4.102267	−1.00408	−0.05326	−1.49206	−0.21214
−1.38515	−1.20449	1.450805	4.730447	−1.24314	−0.16686	−1.70049	−0.25973
−0.5144	−1.01419	1.028582	3.413316	−0.47505	0.085289	−1.40035	−0.22134
−0.89787	−1.1461	1.31354	4.000451	−0.62122	0.083589	−1.57685	−0.27682
−0.82413	−1.14667	1.314847	4.061095	−0.66596	0.005523	−1.57764	−0.3045
−0.61264	−1.30422	1.700983	4.678758	−0.85957	−0.02727	−1.7944	−0.24862
−0.93121	−1.24413	1.547863	4.561717	−0.83257	−0.1856	−1.70556	−0.27807
−1.22028	−1.31118	1.719192	4.694737	−0.91334	0.093265	−1.81818	−0.28616
−1.11261	−1.41217	1.994234	5.232784	−1.13323	0.03697	−1.95823	−0.32969
−1.18002	−1.27892	1.635634	4.804433	−1.10558	−0.18239	−1.76703	−0.32188
−1.0529	−1.16388	1.354619	4.425533	−0.99749	−0.04975	−1.60808	−0.23066
−1.07471	−1.07827	1.16267	4.185393	−1.02387	−0.19871	−1.4898	−0.22694
−1.00023	−0.89588	0.802609	3.319684	−0.71653	0.076915	−1.24691	−0.19552
−1.01639	−1.13042	1.27784	4.32997	−1.03582	0.004731	−1.57333	−0.26091
−1.38288	−1.28598	1.653745	4.991628	−1.27234	−0.0358	−1.78985	−0.31667
−1.39137	−1.3046	1.701985	5.123603	−1.23946	−0.06574	−1.84815	−0.26187
−1.5405	−1.3561	1.839011	5.433238	−1.39961	−0.18786	−1.92762	−0.29242
−31.55	**−36.49**	**45.274**	**134.43**	**−28.66**	**−0.961**	**−51**	**−8.215**

142

14.4 Human Energy Consumed for Face Drilling (Independent Terms)-4

ZB	B	AB	BB	BC	BD	BE	BF
−1.68506	−3.28638	4.623443	10.80027	−1.53935	0.276369	−4.55362	−0.71723
−2.66744	−3.4325	4.946111	11.78209	−1.86051	0.250346	−4.7561	−0.82906
−2.46738	−3.4905	5.488704	12.18357	−2.02721	0.016813	−4.83645	−0.92692
−1.72373	−3.54165	2.894738	12.54328	−2.3342	−0.07406	−4.92433	−0.67512
−2.66724	−3.58741	3.771382	12.86949	−2.40068	−0.53517	−4.98795	−0.80182
−3.30806	−3.50136	3.799374	12.25954	−2.43898	0.249054	−4.92361	−0.76415
−2.78486	−3.64749	3.673473	13.30419	−2.92702	0.09549	−5.1291	−0.85156
−3.50394	−3.70548	4.299493	13.7306	−3.20325	−0.52844	−5.22824	−0.93259
−3.42778	−3.75664	4.666646	14.11231	−3.21957	−0.16059	−5.30041	−0.74448
−3.79297	−3.80239	4.814763	14.45819	−3.61054	−0.70073	−5.36497	−0.80027
−3.03077	−3.6263	5.498376	13.15006	−2.90032	0.311332	−5.18881	−0.79142
−3.47486	−3.77243	4.634773	14.23123	−3.45675	0.015788	−5.3979	−0.87071
−4.15097	−3.83042	5.229974	14.67213	−3.7898	−0.10663	−5.48088	−0.94324
−4.10608	−3.88157	4.102267	15.06662	−3.68775	−0.19559	−5.47996	−0.77913
−4.51639	−3.92733	4.730447	15.42393	−4.05333	−0.54405	−5.54456	−0.84687
−1.70703	−3.36556	3.413316	11.32698	−1.57644	0.283028	−4.64701	−0.73451
−2.73451	−3.4905	4.000451	12.18357	−1.89195	0.254575	−4.80239	−0.84307
−2.54546	−3.54165	4.061095	12.54328	−2.05692	0.017059	−4.87277	−0.9405
−1.68514	−3.58741	4.678758	12.86949	−2.36436	−0.07502	−4.93572	−0.68385
−2.74438	−3.66659	4.561717	13.44387	−2.45367	−0.54699	−5.02646	−0.81951
−3.33232	−3.58054	4.694737	12.8203	−2.49414	0.254686	−4.96506	−0.78143
−2.91943	−3.70548	5.232784	13.7306	−2.97355	0.097009	−5.13831	−0.8651
−3.46613	−3.75664	4.804433	14.11231	−3.24747	−0.53573	−5.19038	−0.94546
−3.43982	−3.80239	4.425533	14.45819	−3.25879	−0.16255	−5.2536	−0.75355
−3.86877	−3.88157	4.185393	15.06662	−3.68572	−0.71532	−5.363	−0.81694
−4.13706	−3.70548	3.319684	13.7306	−2.96365	0.31813	−5.15735	−0.8087
−3.44404	−3.83042	4.32997	14.67213	−3.50989	0.016031	−5.33125	−0.88409
−4.17405	−3.88157	4.991628	15.06662	−3.84041	−0.10805	−5.40244	−0.95584
−4.18853	−3.92733	5.123603	15.42393	−3.73122	−0.1979	−5.56362	−0.78831
−4.55131	−4.00651	5.433238	16.05215	−4.13505	−0.55502	−5.69502	−0.86395
−96.25	**−110.5**	**134.43**	**408.09**	**−87.63**	**−3.286**	**−154.4**	**−24.76**

14.5 Human Energy Consumed for Face Drilling (Independent Terms)-5

ZC	C	AC	BC	CC	CD	CE	CF
0.24017	0.468405	−0.65898	−1.53935	0.219403	−0.03939	0.649025	0.102227
0.421216	0.542028	−0.78104	−1.86051	0.293794	−0.03953	0.751037	0.130917
0.410545	0.58078	−0.91326	−2.02721	0.337306	−0.0028	0.804733	0.154229
0.320773	0.659072	−0.53869	−2.3342	0.434376	0.013783	0.916377	0.125635
0.497547	0.669197	−0.70352	−2.40068	0.447824	0.099832	0.930454	0.149571
0.658125	0.696581	−0.75587	−2.43898	0.485224	−0.04955	0.979531	0.152025
0.612689	0.802474	−0.80819	−2.92702	0.643964	−0.02101	1.128438	0.187349
0.817442	0.864461	−1.00304	−3.20325	0.747294	0.123281	1.219708	0.217566
0.78201	0.857035	−1.06464	−3.21957	0.73451	0.036637	1.20923	0.169845
0.94719	0.949544	−1.20236	−3.61054	0.901633	0.174989	1.339755	0.199846
0.668452	0.7998	−1.2127	−2.90032	0.63968	−0.06867	1.144419	0.174552
0.84404	0.916318	−1.12578	−3.45675	0.839639	−0.00384	1.311142	0.211493
1.072193	0.989396	−1.3509	−3.7898	0.978904	0.027542	1.415708	0.243638
1.005016	0.950066	−1.00408	−3.68775	0.902626	0.047874	1.341292	0.190702
1.186883	1.032082	−1.24314	−4.05333	1.065194	0.142973	1.457081	0.222554
0.237578	0.468405	−0.47505	−1.57644	0.219403	−0.03939	0.646752	0.102227
0.424633	0.542028	−0.62122	−1.89195	0.293794	−0.03953	0.745747	0.130917
0.417419	0.58078	−0.66596	−2.05692	0.337306	−0.0028	0.799064	0.154229
0.309591	0.659072	−0.85957	−2.36436	0.434376	0.013783	0.906782	0.125635
0.500882	0.669197	−0.83257	−2.45367	0.447824	0.099832	0.91739	0.149571
0.64829	0.696581	−0.91334	−2.49414	0.485224	−0.04955	0.965932	0.152025
0.632244	0.802474	−1.13323	−2.97355	0.643964	−0.02101	1.112772	0.187349
0.797611	0.864461	−1.10558	−3.24747	0.747294	0.123281	1.194389	0.217566
0.775313	0.857035	−0.99749	−3.25879	0.73451	0.036637	1.184128	0.169845
0.94641	0.949544	−1.02387	−3.68572	0.901633	0.174989	1.311943	0.199846
0.892952	0.7998	−0.71653	−2.96365	0.63968	−0.06867	1.113175	0.174552
0.823888	0.916318	−1.03582	−3.50989	0.839639	−0.00384	1.275347	0.211493
1.063946	0.989396	−1.27234	−3.84041	0.978904	0.027542	1.377058	0.243638
1.013252	0.950066	−1.23946	−3.73122	0.902626	0.047874	1.345902	0.190702
1.172422	1.032082	−1.39961	−4.13505	1.065194	0.142973	1.467042	0.222554
21.141	**23.554**	**−28.66**	**−87.63**	**19.343**	**0.8843**	**32.961**	**5.2643**

14.6 Human Energy Consumed for Face Drilling (Independent Terms)-6

ZD	D	AD	BD	CD	DD	DE	DF
−0.04312	−0.0841	0.11831	0.276369	−0.03939	0.007072	−0.11652	−0.01835
−0.05668	−0.07293	0.105095	0.250346	−0.03953	0.005319	−0.10106	−0.01762
−0.0034	−0.00482	0.007574	0.016813	−0.0028	2.32E−05	−0.00667	−0.00128
0.010178	0.020912	−0.01709	−0.07406	0.013783	0.000437	0.029077	0.003986
0.110916	0.149181	−0.15683	−0.53517	0.099832	0.022255	0.207422	0.033343
−0.0672	−0.07113	0.077185	0.249054	−0.04955	0.00506	−0.10002	−0.01552
−0.01999	−0.02618	0.026366	0.09549	−0.02101	0.000685	−0.03681	−0.00611
0.134853	0.14261	−0.16547	−0.52844	0.123281	0.020338	0.201215	0.035892
0.039006	0.042749	−0.0531	−0.16059	0.036637	0.001827	0.060316	0.008472
0.18383	0.184287	−0.23335	−0.70073	0.174989	0.033962	0.260019	0.038786
−0.07175	−0.08585	0.130176	0.311332	−0.06867	0.007371	−0.12285	−0.01874
−0.00386	−0.00419	0.005142	0.015788	−0.00384	1.75E−05	−0.00599	−0.00097
0.030167	0.027838	−0.03801	−0.10663	0.027542	0.000775	0.039832	0.006855
0.053305	0.05039	−0.05326	−0.19559	0.047874	0.002539	0.07114	0.010115
0.159306	0.138528	−0.16686	−0.54405	0.142973	0.01919	0.195573	0.029872
−0.04265	−0.0841	0.085289	0.283028	−0.03939	0.007072	−0.11612	−0.01835
−0.05714	−0.07293	0.083589	0.254575	−0.03953	0.005319	−0.10035	−0.01762
−0.00346	−0.00482	0.005523	0.017059	−0.0028	2.32E−05	−0.00663	−0.00128
0.009823	0.020912	−0.02727	−0.07502	0.013783	0.000437	0.028772	0.003986
0.111659	0.149181	−0.1856	−0.54699	0.099832	0.022255	0.20451	0.033343
−0.0662	−0.07113	0.093265	0.254686	−0.04955	0.00506	−0.09864	−0.01552
−0.02063	−0.02618	0.03697	0.097009	−0.02101	0.000685	−0.0363	−0.00611
0.131582	0.14261	−0.18239	−0.53573	0.123281	0.020338	0.197038	0.035892
0.038672	0.042749	−0.04975	−0.16255	0.036637	0.001827	0.059064	0.008472
0.183679	0.184287	−0.19871	−0.71532	0.174989	0.033962	0.254622	0.038786
−0.09585	−0.08585	0.076915	0.31813	−0.06867	0.007371	−0.11949	−0.01874
−0.00376	−0.00419	0.004731	0.016031	−0.00384	1.75E−05	−0.00583	−0.00097
0.029935	0.027838	−0.0358	−0.10805	0.027542	0.000775	0.038745	0.006855
0.053741	0.05039	−0.06574	−0.1979	0.047874	0.002539	0.071385	0.010115
0.157365	0.138528	−0.18786	−0.55502	0.142973	0.01919	0.19691	0.029872
0.8823	**0.8146**	**−0.961**	**−3.286**	**0.8843**	**0.2537**	**1.1424**	**0.1775**

14.7 Human Energy Consumed for Face Drilling (Independent Terms)-7

ZE	E	AE	BE	CE	DE	EE	EF
0.710455	1.385606	−1.94934	−4.55362	0.649025	−0.11652	1.919905	0.302401
1.076771	1.385606	−1.99661	−4.7561	0.751037	−0.10106	1.919905	0.334668
0.979464	1.385606	−2.17883	−4.83645	0.804733	−0.00667	1.919905	0.367954
0.676715	1.390405	−1.13644	−4.92433	0.916377	0.029077	1.933226	0.265045
1.033766	1.390405	−1.46171	−4.98795	0.930454	0.207422	1.933226	0.310767
1.328568	1.406199	−1.52589	−4.92361	0.979531	−0.10002	1.977397	0.306895
1.073634	1.406199	−1.41622	−5.1291	1.128438	−0.03681	1.977397	0.328297
1.334202	1.410946	−1.63713	−5.22824	1.219708	0.201215	1.990768	0.355105
1.287431	1.410946	−1.75273	−5.30041	1.20923	0.060316	1.990768	0.279618
1.407448	1.410946	−1.7866	−5.36497	1.339755	0.260019	1.990768	0.296956
1.195892	1.430881	−2.16957	−5.18881	1.144419	−0.12285	2.04742	0.312281
1.318014	1.430881	−1.75797	−5.3979	1.311142	−0.00599	2.04742	0.330258
1.550623	1.430881	−1.95369	−5.48088	1.415708	0.039832	2.04742	0.352354
1.493443	1.411788	−1.49206	−5.47996	1.341292	0.07114	1.993145	0.283381
1.623541	1.411788	−1.70049	−5.54456	1.457081	0.195573	1.993145	0.304432
0.700327	1.380754	−1.40035	−4.64701	0.646752	−0.11612	1.906481	0.301341
1.077861	1.375846	−1.57685	−4.80239	0.745747	−0.10035	1.892953	0.33231
0.988849	1.375846	−1.57764	−4.87277	0.799064	−0.00663	1.892953	0.365363
0.646287	1.375846	−1.7944	−4.93572	0.906782	0.028772	1.892953	0.26227
1.026081	1.370883	−1.70556	−5.02646	0.91739	0.20451	1.87932	0.306404
1.290546	1.386677	−1.81818	−4.96506	0.965932	−0.09864	1.922874	0.302634
1.09252	1.386677	−1.95823	−5.13831	1.112772	−0.0363	1.922874	0.323739
1.274811	1.381656	−1.76703	−5.19038	1.194389	0.197038	1.908975	0.347733
1.249909	1.381656	−1.60808	−5.2536	1.184128	0.059064	1.908975	0.273813
1.377097	1.381656	−1.4898	−5.363	1.311943	0.254622	1.908975	0.290791
1.553921	1.391817	−1.24691	−5.15735	1.113175	−0.11949	1.937154	0.303756
1.251423	1.391817	−1.57333	−5.33125	1.275347	−0.00583	1.937154	0.321242
1.496688	1.391817	−1.78985	−5.40244	1.377058	0.038745	1.937154	0.342734
1.510857	1.416641	−1.84815	−5.56362	1.345902	0.071385	2.00687	0.284355
1.614723	1.421439	−1.92762	−5.69502	1.467042	0.19691	2.02049	0.306513
36.242	**41.91**	**−51**	**−154.4**	**32.961**	**1.1424**	**58.558**	**9.3954**

14.8 Human Energy Consumed for Face Drilling (Independent Terms)-8

ZF	F	AF	BF	CF	DF	EF	FF
0.111902	0.218244	−0.30704	−0.71723	0.102227	−0.01835	0.302401	0.047631
0.187697	0.241532	−0.34804	−0.82906	0.130917	−0.01762	0.334668	0.058337
0.187717	0.265555	−0.41758	−0.92692	0.154229	−0.00128	0.367954	0.070519
0.092778	0.190624	−0.15581	−0.67512	0.125635	0.003986	0.265045	0.036338
0.166178	0.223508	−0.23497	−0.80182	0.149571	0.033343	0.310767	0.049956
0.206196	0.218244	−0.23682	−0.76415	0.152025	−0.01552	0.306895	0.047631
0.17825	0.233464	−0.23513	−0.85156	0.187349	−0.00611	0.328297	0.054505
0.237989	0.251678	−0.29202	−0.93259	0.217566	0.035892	0.355105	0.063342
0.180829	0.198178	−0.24618	−0.74448	0.169845	0.008472	0.279618	0.039274
0.209944	0.210466	−0.2665	−0.80027	0.199846	0.038786	0.296956	0.044296
0.182403	0.218244	−0.33091	−0.79142	0.174552	−0.01874	0.312281	0.047631
0.212602	0.230808	−0.28357	−0.87071	0.211493	−0.00097	0.330258	0.053272
0.266857	0.24625	−0.33622	−0.94324	0.243638	0.006855	0.352354	0.060639
0.212334	0.200725	−0.21214	−0.77913	0.190702	0.010115	0.283381	0.04029
0.247979	0.215636	−0.25973	−0.84687	0.222554	0.029872	0.304432	0.046499
0.110695	0.218244	−0.22134	−0.73451	0.102227	−0.01835	0.301341	0.047631
0.18922	0.241532	−0.27682	−0.84307	0.130917	−0.01762	0.33231	0.058337
0.19086	0.265555	−0.3045	−0.9405	0.154229	−0.00128	0.365363	0.070519
0.089544	0.190624	−0.24862	−0.68385	0.125635	0.003986	0.26227	0.036338
0.167292	0.223508	−0.27807	−0.81951	0.149571	0.033343	0.306404	0.049956
0.203114	0.218244	−0.28616	−0.78143	0.152025	−0.01552	0.302634	0.047631
0.183939	0.233464	−0.32969	−0.8651	0.187349	−0.00611	0.323739	0.054505
0.232216	0.251678	−0.32188	−0.94546	0.217566	0.035892	0.347733	0.063342
0.17928	0.198178	−0.23066	−0.75355	0.169845	0.008472	0.273813	0.039274
0.209771	0.210466	−0.22694	−0.81694	0.199846	0.038786	0.290791	0.044296
0.243663	0.218244	−0.19552	−0.8087	0.174552	−0.01874	0.303756	0.047631
0.207526	0.230808	−0.26091	−0.88409	0.211493	−0.00097	0.321242	0.053272
0.264804	0.24625	−0.31667	−0.95584	0.243638	0.006855	0.342734	0.060639
0.214074	0.200725	−0.26187	−0.78831	0.190702	0.010115	0.284355	0.04029
0.244957	0.215636	−0.29242	−0.86395	0.222554	0.029872	0.306513	0.046499
5.8126	**6.7263**	**−8.215**	**−24.76**	**5.2643**	**0.1775**	**9.3954**	**1.5203**

14.9 Error Calculation in Human Energy for Face Drilling

Cal T	Exp T = pD3 = He/ Dr3 * So	Error = ET − CT	% Error
3.169914	3.256415279	0.0865	2.656347376
4.582779	5.985657551	1.40288	23.43734088
5.348908	5.091961407	0.25695	5.046127848
4.685353	3.066925439	1.61843	52.77035742
5.100239	5.539869779	0.43963	7.935766564
5.750254	8.806299006	3.05604	34.70294196
8.118918	5.800968168	2.31795	39.95798243
8.689694	8.822842891	0.13315	1.509134135
7.784008	8.174471432	0.39046	4.77662414
8.449815	9.943080914	1.49327	15.01813878
7.737109	6.851307189	0.8858	12.92894702
10.80105	8.33913453	2.46192	29.52246245
12.93976	12.12506649	0.81469	6.719050474
10.41725	11.42452046	1.00727	8.816738302
11.72593	14.12501798	2.39909	16.98465574
3.648314	3.215186515	0.43313	13.47131368
5.008813	6.073187588	1.06437	17.52579217
5.879263	5.232639012	0.64662	12.35750934
4.554549	2.949429224	1.60512	54.42138234
5.301225	5.603792085	0.30257	5.399320183
5.954268	8.52462096	2.57035	30.15210886
8.003728	6.135766627	1.86796	30.44380912
8.565934	8.368898407	0.19704	2.354377132
7.790159	8.028701089	0.23854	2.971122993
8.960044	9.924303008	0.96426	9.716135532
8.419134	13.07584267	4.65671	35.61306451
10.76019	7.927364829	2.83282	35.73471084
12.74436	11.89456814	0.84979	7.144386164
10.90205	11.65486625	0.75281	6.459230725
13.13275	13.67657965	0.54383	3.976350984

Index

Printed in the United States
by Baker & Taylor Publisher Services